W9-AED-855

Lectures in Applied Mathematics

Proceedings of the Summer Seminar, Boulder, Colorado, 1960

VOLUME I   LECTURES IN STATISTICAL MECHANICS
            G. E. Uhlenbeck and G. W. Ford with E. W. Montroll

VOLUME II   MATHEMATICAL PROBLEMS OF RELATIVISTIC PHYSICS
            I. E. Segal with G. W. Mackey

VOLUME III PERTURBATION OF SPECTRA IN HILBERT SPACE
            K. O. Friedrichs

VOLUME IV   QUANTUM MECHANICS
            V. Bargmann

Proceedings of the Summer Seminar, Ithaca, New York, 1963

VOLUME V   SPACE MATHEMATICS, PART 1
            J. Barkley Rosser, Editor

VOLUME VI   SPACE MATHEMATICS, PART 2
            J. Barkley Rosser, Editor

VOLUME VII SPACE MATHEMATICS, PART 3
            J. Barkley Rosser, Editor

Volume 5
Lectures in Applied Mathematics

SPACE MATHEMATICS
PART 1

J. Barkley Rosser, EDITOR
MATHEMATICS RESEARCH CENTER
THE UNIVERSITY OF WISCONSIN

1966
AMERICAN MATHEMATICAL SOCIETY, PROVIDENCE, RHODE ISLAND

CARL A. RUDISILL LIBRARY
LENOIR RHYNE COLLEGE

6 29.411
Su 65
5 7906
May 1968

Supported by the

National Aeronautics and Space Administration under Research Grant NsG 358

Air Force Office of Scientific Research under Grant AF-AFOSR 258-63

Army Research Office (Durham) under Contract DA-31-124-ARO(D)-82

Atomic Energy Commission under Contract AT(30-1)-3164

Office of Naval Research under Contract Nonr(G)00025-63

National Science Foundation under NSF Grant GE-2234

All rights reserved except those granted to the United States Government, otherwise, this book, or parts thereof, may not be reproduced in any form without permission of the publishers.

Library of Congress Catalog Card Number 66-20435

Copyright © 1966 by the American Mathematical Society

Printed in the United States of America

CARL A. RUDISILL LIBRARY
LENOIR RHYNE COLLEGE

Contents

v

Part 2

Part 3

Foreword

In 1962 a committee of the Division of Mathematics of the National Research Council presented to the Space Science Board of the National Academy of Sciences a report on the current and anticipated uses of mathematics in space activities. A key finding was the need for much more intensive work in various mathematical areas, both to develop more powerful results in these areas and to discover ways of using the existing knowledge more effectively. The Summer Seminar of 1963 was planned to fill this need in part. It was the third Summer Seminar in Applied Mathematics sponsored by the American Mathematical Society.

In order to keep the Seminar within bounds and to achieve reasonable coherence, many mathematical areas of great importance and urgency for space activity were not considered at the Seminar. The most notable omissions are the area of nonlinear differential equations, which is of use in the study of guidance systems for propulsive space vehicles, and various areas of statistics, such as those involved in the design of experiments to be performed in space, analysis of large amounts of data from experiments aboard space vehicles, etc.

Primarily, the topics considered at the Seminar were those having to do with the behavior of nonpropulsive space vehicles. Inevitably, this led to heavy emphasis on the theory of orbits. For this reason, the Seminar was combined with the fifth annual Summer Institute in Dynamical Astronomy. Four previous Summer Institutes in Dynamical Astronomy have been sponsored by Yale University, in the summers of 1959 through 1962. They were devoted primarily to topics in the theory of orbits, and were more restricted in scope than the present Seminar.

The orbits of most present and planned space vehicles differ so markedly from the orbits of the classical celestial bodies that only parts of the classical theories of celestial mechanics are applicable. Much effort is being expended to determine which of the classical methods are applicable, to find suitable modifications of some of the classical methods to make them more widely applicable, and to find new methods to handle novel situations.

Such matters were a main concern of the four previous Summer Institutes in Dynamical Astronomy, and received a considerable amount of attention at the present Seminar; it was to assure this that the Seminar was combined with the fifth Institute. Notes were compiled for the Institute of 1959, but their supply is exhausted. Notes for the Institute of 1960 were edited by William E. Felling and published in 1961 by The McDonnell Aircraft Corporation, St. Louis, Missouri, under the title, "Notes of the Summer Institute in Dynamical Astronomy at Yale University, July 1960". Copies of these notes can be obtained from The McDonnell Aircraft Corporation.

The Institute of 1961 was held at Tucson, Arizona, and the Institute of 1962 was held at Yale University. Only scattered notes, mostly in mimeographed form, survive from these meetings. Some of these were collected and have been included in the present Proceedings. Specifically, from the Institute of 1961 come the chapters: *Precession and nutation* by Alan Fletcher; *Lectures on regularization* by Paul B. Richards. From the Institute of 1962 come the chapters: *Problems of stellar dynamics* by G. Contopoulos; *Matrix methods* by J.M.A. Danby; *The effect of radiation pressure on the motion of an artificial satellite* by Gen-ichiro Hori; *The two variable expansion procedure for the approximate solution of certain nonlinear differential equations* by J. Kevorkian; *Stability and small oscillations about equilibrium and periodic motions* by P. J. Message; *The spheroidal method in satellite astronomy* by John P. Vinti.

Some of the lectures of the earlier Institutes, and of the present Seminar, have been published elsewhere, either separately or as parts of larger works. Indeed, some lectures excerpted particularly relevant material already in print. Besides the Notes of the 1960 Institute already cited, a list of publications containing material not covered in the present Proceedings which was presented in lectures at the Institutes and the Seminar follows:

R. F. Arenstorf, *Periodic solutions of the restri~ted three body problem presenting analytic continuations of Keplerian motions,* Amer. J. Math 85(1963), 27-35.

V. I. Arnol'd, *The stability of the equilibrium position of a Hamiltonian system of ordinary differential equations in the general elliptic case,* Dokl. Akad. Nauk SSSR **137**(1961), 255-257 = Soviet Math. Dokl. **2**(1961), 247.

_____, *Generation of quasi-periodic motion from a family of periodic motions,* Dokl. Akad. Nauk SSSR **138**(1961), 13-15 = Soviet Math. Dokl. **2**(1961), 501.

_____, *The classical theory of perturbations and the problem of stability of planetary systems,* Dokl. Akad. Nauk SSSR **145**(1962), 487-490 = Soviet Math. Dokl. **3**(1962), 1008.

_____, *Proof of A. N. Kolmogorov's theorem on the preservation of quasi-periodic motions under small perturbations of the Hamiltonian,* Uspehi Mat. Nauk SSSR **18**, Ser. 5 (113), 1963, pp. 13-40.

_____, *Small divisor and stability problems in classical and celestial mechanics,* Uspehi Mat. Nauk SSSR **18**, Ser. 6(119), 1963, pp. 81-192.

R. E. Bellman, *Adaptive control processes; a guided tour,* Princeton Univ. Press, Princeton, N. J., 1961.

R. E. Bellman and S. Dreyfus, *Applied dynamic programming,* Princeton Univ. Press, Princeton, N. J., 1962.

D. Brouwer, *Solution of the problem of artificial satellite motion without drag,* Astronom. J. **64**(1959), 378-397.

D. Brouwer and G. M. Clemence, *Methods of celestial mechanics,* Academic Press, New York, 1961.

D. Brouwer and G. Hori, *Theoretical evaluation of atmospheric drag effects in the motion of an artificial satellite,* Astronom. J. **66**(1961), 193-225; 264-265.

C. J. Cohen and E. C. Hubbard, *A non-singular set of orbit elements,* Astronom. J. **67**(1962), 10-15.

C. C. Conley, *A disk mapping associated with the satellite problem,*

Comm. Pure Appl. Math. **17**(1964), 237-243.

S. Dreyfus, *Dynamic programming and the calculus of variations,* J. Math. Anal. Appl. **1**(1960), 228-239.

W. J. Eckert and D. Brouwer, *The use of rectangular coordinates in the differential correction of orbits,* Astronom. J. **46**(1937), 125-132.

B. Garfinkel, *Motion of a satellite in the vicinity of the critical inclination,* Astronom. J. **65**(1960), 624-627.

Y. Hagihara, *Theories of equilibrium figures of a homogeneous rotating fluid mass,* Blaisdell, New York, forthcoming.

———, *Celestial mechanics,* Blaisdell, New York, forthcoming in four volumes.

Paul Herget, *The computation of orbits,* Observatory of the University of Cincinnati, 1948.

———, *Computation of preliminary orbits,* Astronom. J. **70**(1965), 1-2.

G. Hori, *Motion of an artificial satellite in the vicinity of the critical inclination,* Astronom. J. **65**(1960), 291-300.

A. N. Kolmogorov, Dokl. Akad. Nauk SSSR **98**(1954), 527-530.

———, *General theory of dynamical systems and classical mechanics,* Proc. Internat. Congr. Math., Amsterdam, 1954, (Amsterdam: Erven P. Nordhoff, 1957), Vol. 1, pp. 315-333.

W. T. Kyner, *Qualitative properties of orbits about an oblate planet,* Comm. Pure Appl. Math. **17**(1964), 227-236.

———, *A mathematical theory of the orbits about an oblate planet,* J. Soc. Indust. Appl. Math. **13**(1965), 136-171.

J. Moser, *Nonexistence of integrals for canonical systems of differential equations,* Comm. Pure Appl. Math. **8**(1955), 409-436.

———, *On invariant curves of area-preserving mappings of an annulus,* Nachr. Akad. Wiss. Göttingen Math.-Phys. Kl., No. 1 (1962), 1-20.

———, *Stability and nonlinear character of ordinary differential equations,* Proc. Sympos. on Nonlinear problems, 1962, Univ. of Wisconsin Press, Madison, Wis., 1963, pp. 139-150.

P. Musen, *Special perturbations of the vectorial elements,* Astronom. J. **59**(1954), 262-267.

S. Ostrach, *Melting ablation,* to appear in 1966 as one of the International Series of Monographs on Interdisciplinary and Advanced Topics in Science and Engineering, Pergamon Press, Oxford.

C. L. Siegel, *Iteration of analytic functions,* Ann. of Math. **43**(1942), 607-612.

_____, *Über die Existenz einer Normalform analytischer Hamilton-scher Differentialgleichungen in der Nähe einer Gleichgewichtlösung,* Math. Ann. **128**(1954), 144-170.

Together with the above references, the present Proceedings should acquaint the reader with the current state of research on the behavior of nonpropulsive space vehicles, indicate the more pressing unsolved problems, and furnish examples of mathematical techniques which are currently useful. It is hoped that the reader will be stimulated to make contributions of his own, either in the way of developing better mathematical techniques, or finding more ingenious uses of existing ones.

Besides presenting much new and advanced material, an effort is made in these Proceedings to give readers basic information, in fields other than their own, which they need to have a full understanding of space problems in their own fields. Accordingly an effort is made to acquaint the mathematical specialists with the key space problems in aerodynamics, geophysics, orbit theory, etc.; to acquaint orbit specialists with useful mathematical techniques, and to give them enough background in geophysics, aerodynamics, etc. for them to see the relevance of these areas for the new orbits required for space exploration; and so on, for other areas represented in space activity. Thus, for the benefit of the readers who are not specialists in orbit theory, there is included the chapter *Elliptic motion* by J. M. A. Danby. This is a summary exposition of the simplest orbit problem, the problem of two bodies. However, it includes only those points of especial relevance for space activity, and makes no effort to cover all developments that have been made in the two-body problem in the more than three hundred years since Newton gave the first solution.

Useful mathematical techniques for orbit studies are presented in the chapters *Matrix methods* by J. M. A. Danby, *The Lagrange-Hamilton-Jacobi mechanics* by Boris Garfinkel, *Stability and small oscillations about equilibrium and periodic motions* by P. J. Message, *Lectures on regularization* by Paul B. Richards, and *The spheroidal method in satellite astronomy* by John P. Vinti. For near earth satellites, precession and nutation play such a prominent role that the chapter *Precession and nutation* by Alan Fletcher is almost required reading for anyone concerned with near earth satellites. For satellites that go further afield and also come near the moon, the motion is essentially covered by the "restricted problem of three bodies" in which the mass of one body is infinitesimal compared to the masses of the other two. This is dealt with in the chapter *Notes on a two-degree-of-freedom irreversible dynamical system: the restricted problem of three bodies* by Victor Szebehely. The chapters *Problems of stellar dynamics* by George Contopoulos and *Qualitative methods in the n-body problem* by Harry Pollard are explanations of some classic techniques of interest.

To a first approximation, a small object will remain at one vertex of an equilateral triangle of which the earth and moon are the other two vertices; it rotates at the same rate as the earth and moon about a common center of gravity. This "triangular point" has been proposed as a good place to park a permanent space station, say for communication purposes. However, the pull of the sun would cause the space station to wander about the "triangular point," perhaps to the extent of rendering it useless. Though this question is of immediate concern, no completely satisfactory treatment exists. Five chapters deal with aspects of this, namely *Motion in the vicinity of the triangular libration centers* by André Deprit, *Motion of a particle in the vicinity of a triangular libration point in the earth-moon system* by J. Pieter de Vries, *The dominant features of the long-period librations of the Trojan minor planets* by P. J. Message, *Outline of a theory of nonperiodic motions in the neighborhood of the long-period librations about the equilateral points of the restricted problem of three bodies* by Eugene Rabe, and *Elements of a theory of librational motions in the elliptical restricted problem* by Eugene Rabe.

Many artificial satellites have orbits very close to the earth, and as a result these orbits are much more perturbed by irregularities in the potential field of the earth than is the case for the more distant natural celestial objects. This is compelling a more careful study of the distribution of masses within the earth, as well as requiring new methods for taking account of the effects on orbits of irregularities in this distribution. Thus it is that two chapters are devoted to questions of geophysics, and relevant matters, namely, *The shape of the earth in the light of recent discoveries in space science* by John A. O'Keefe and *The stability of a rotating liquid mass* by John A. O'Keefe.

If the orbit of some satellite is known with sufficient precision over a considerable period of time, one can make many inferences about the potential field of the earth. The determination of such an orbit is based on observations from various stations on the earth. To deter-

mine the orbit with precision from these observations requires knowledge of the positions of the stations relative to the equipotential
surfaces of the earth's potential field, and hence some knowledge of
the potential field. The question of finding what orbits and potential
fields are consistent with a set of observations is very difficult at best.
Since the observations cannot be totally free from error, it may be
that no orbit or potential field is exactly consistent with them, and
one must use variational and optimization techniques to find an orbit
and potential field that gives a "best fit." A specific discussion of this
point is given in the chapter *Geodetic problems and satellite orbits*
by W. H. Guier. Relevant general mathematical techniques are
discussed in the chapter *Elements of calculus of variations and opti-
mum control theory* by Magnus R. Hestenes.

Many artificial satellites have low orbits, where the drag of the
atmosphere is not negligible. To understand better the nature of this
drag requires investigation into some difficult areas of aerodynamics.
Presentations were made at the Seminar of the latest results in this
area, and are recorded in the chapters *Basic fluid dynamics* by S. F.
Shen and *Shock waves in rarefied gases* by S. F. Shen. In extreme
cases, such as re-entry or flow inside rocket nozzles, there may be
chemical dissociation and other effects. These are discussed in the
chapter *Models of gas flows with chemical and radiative effects* by
F. K. Moore. The specific effects of drag on a satellite orbit are discussed in the chapter *Decay of orbits* by P. J. Message. Other
peculiarities of the orbits of artificial satellites are discussed in the
chapter *The effect of radiation pressure on the motion of an artificial
satellite* by Gen-ichiro Hori.

In all too many problems of the most immediate urgency, no
suitable theoretical treatments have yet been perfected, and the best
that can be done is to seek approximate numerical answers by means
of high speed computers. A discussion of computational techniques
is given in the chapter *Special computation procedures for differential
equations* by S. V. Parter.

Although very little was said at the 1963 Seminar about general
techniques for solving nonlinear differential equations, a chapter
on the subject is included from an earlier Institute, namely *The two
variable expansion procedure for the approximate solution of certain
nonlinear differential equations* by J. Kevorkian.

A topic which now particularly timely, namely the rendezvous of

two space vehicles, is treated in *Rendezvous problems* by J. C. Houbolt.
The Seminar was supported by the following government agencies:

> National Science Foundation,
> Air Force Office of Scientific Research,
> National Aeronautics and Space Administration,
> Army Research Office (Durham),
> Office of Naval Research,
> Atomic Energy Commision.

The success of such a Seminar is the result of much dedicated work
and planning by many individuals. Gordon L. Walker, Executive
Director of the American Mathematical Society, contributed greatly
to the planning and functioning of the Seminar. Edmund T. Cranch,
of the Engineering College at Cornell University, served as Adminis-
trative Director. He handled matters of preparation skillfully and,
with the assistance of Margaret Kellar from the Office of the
American Mathematical Society, managed the day-by-day operation
of the Seminar in a highly satisfactory fashion. The enthusiastic
office staff of Ann Smith, Virginia Cranch, and Neva Strever were
effective in the many things, both obvious and unseen, that they did
to promote the success and unity of the Seminar. Space for offices
and meetings, and many forms of cooperation, were furnished by
Cornell University and members of its staff. The scientific planning
and direction were done by the Joint Invitations and Organizing
Committee. Its members were Dirk Brouwer, Director of Yale
University Observatory; Morris Davis, Director of Yale Computer
Center; William R. Sears, Director of the Graduate School of Aero-
space Engineering of Cornell University; Victor Szebehely, Manager
of Space Mechanics Operation, General Electric Company; and
Chairman, J. Barkley Rosser, Director U. S. Army Mathematics
Research Center, The University of Wisconsin. During the Seminar,
Dirk Brouwer served as Associate Director and Morris Davis as
Assistant Director. They assumed day-by-day responsibility for the
staff and program, and also furnished valuable advice and continuity
with previous Institutes.

J. BARKLEY ROSSER

J. M. A. Danby

Elliptic Motion

---

1. Introduction to the mechanics of celestial bodies.

1.1. *Newton's Law of Gravitation*. Celestial mechanics is, in the main, a branch of Newtonian mechanics, and the fundamental law is Newton's law of gravitation. It is true that this law may not cover every contingency in cosmogony, but its inadequacies in celestial mechanics are few indeed. Also, in cases where the relevant arguments of general relativity have achieved explicit forms, the resulting modifications to motion governed by Newton's laws have been dealt with by established perturbation theories of celestial mechanics (see [5]).

Newton's law states that: "any two particles attract each other with a force that is proportional to the product of their masses and inversely proportional to the square of the distance between them". Let the particles have masses m_1 and m_2, and position vectors $\mathbf{r}_1$ and $\mathbf{r}_2$, respectively. Then the force exerted by m_2 on m_1 can be written

$$- k^2 m_1 m_2 (\mathbf{r}_1 - \mathbf{r}_2) / |\mathbf{r}_1 - \mathbf{r}_2|^3.$$

(In these notes a vector is written in boldface type. A unit vector is written with a "cap" above it, i.e., $\hat{\mathbf{r}}$, and $|\mathbf{r}|$, or simply "r" stand for the modulus of $\mathbf{r}$. It is assumed that the reader is acquainted with

1

elementary vector algebra and calculus; if not, see [2].) k^2 is often written as "G", and the value in c.g.s. units is about 6.67×10^{-8}.

Consider a system of masses $m_1, m_2, \cdots, m_n$ at $\mathbf{r}_1, \mathbf{r}_2, \cdots, \mathbf{r}_n$. They will exert a total force on a mass m at $\mathbf{r}$ of amount

$$- k^2 m \sum_{i=1}^{n} m_i(\mathbf{r} - \mathbf{r}_i) / |\mathbf{r} - \mathbf{r}_i|^3.$$

The particle m will experience some force wherever it is, and the n bodies are said to set up a "field of force". The strength of a field of force at a point $\mathbf{r}$ is the force exerted on a particle of unit mass placed at $\mathbf{r}$. Strictly, the word "force" in this context means "force per unit mass". The n bodies produce, therefore, a field of force

(1.1.1) $$- k^2 \sum_{i=1}^{n} m_i(\mathbf{r} - \mathbf{r}_i) / |\mathbf{r} - \mathbf{r}_i|^3.$$

In the system of n masses the force acting on m_i is

$$- k^2 \sum_{j=1; j \neq i}^{n} m_i m_j(\mathbf{r}_i - \mathbf{r}_j) / |\mathbf{r}_i - \mathbf{r}_j|^3.$$

This can be calculated from the gradient of the "force function"

(1.1.2) $$U = k^2 \sum_{j=1}^{n} \sum_{i<j} m_i m_j / |\mathbf{r}_i - \mathbf{r}_j|.$$

For instance, the x-component would be $\partial U / \partial x_i$. Since

$$|\mathbf{r}_i - \mathbf{r}_j| = \{(x_i - x_j)^2 + (y_i - y_j)^2 + (z_i - z_j)^2\}^{1/2}$$

then

$$\partial |\mathbf{r}_i - \mathbf{r}_j| / \partial x_i = (x_i - x_j)\{(x_i - x_j)^2 + (y_i - y_j)^2 + (z_i - z_j)^2\}^{-1/2}$$

and

$$\partial U / \partial x_i = - k^2 \sum_{j=1; j \neq i}^{n} m_i m_j(x_i - x_j) / |\mathbf{r}_i - \mathbf{r}_j|^3.$$

The force function is the negative of the work that would be done in assembling the system of n bodies from a state of infinite diffusion. As the words are normally used, it is minus the potential; but this convention is not universal, and I shall use only force functions here.

The transition from particles to solid bodies is accomplished by integration. Consider the force function of a uniform, thin spherical

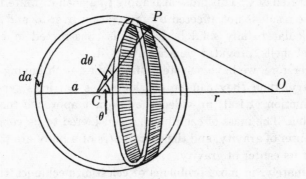

FIGURE 1. Force Due to a Shell

shell at a point O outside the shell (see Figure 1). Let the shell have center C, radius a, thickness da, and density ρ; and let $OC = r$. If P is a point on the shell, let the angle $OCP = \theta$. Divide the shell into thin rings perpendicular to OC and defined by θ lying within the limits θ and $\theta + d\theta$. The radius of a ring is $a \sin\theta$, and its mass is

$$\rho 2\pi a \sin\theta\, a\, d\theta\, da.$$

Any element of the ring is at the distance

$$(r^2 + a^2 - 2ar\cos\theta)^{1/2}$$

from O, so that the force function of the ring at O is

$$k^2 \rho 2\pi a^2\, da \sin\theta d\theta (r^2 + a^2 - 2ar\cos\theta)^{-1/2},$$

and the total force function due to the shell is

$$U = k^2 \rho 2\pi a^2\, da \int_0^\pi \sin\theta d\theta (r^2 + a^2 - 2ar\cos\theta)^{-1/2},$$

where the square root must always be positive. This can be integrated at once to give

$$U = \frac{1}{2}\, k^2\, dm \left[\frac{1}{ra} (r^2 + a^2 - 2ar\cos\theta)^{1/2} \right]_{\theta=0}^\pi$$

$$= k^2\, dm/r,$$

where $dm = 4\pi a^2 \rho\, da$ is the mass of the shell. This means that, so far as O is concerned, the shell could just as well have all its mass concentrated at C. This must also apply to a shell of finite thickness, since the result is not affected by integration over a, and it applies in particular to any solid body that is constructed in concentric spherical shells, provided we are outside it.

If, therefore, we have a system of n bodies, each having spherical symmetry, then they can be considered as particles generating a force function (1.1.2), provided they do not approach too close to each other. The mass of each body is considered to be concentrated at its center of gravity, and the coordinates of a body are the coordinates of its center of gravity.

Fortunately, in most problems of celestial mechanics the bodies can be assumed to be spheres. In the first place they are, in fact, nearly spherical, and in the second place the distances between the bodies are usually large compared with the dimensions of the bodies themselves. In the case of the motion of an artificial satellite the latter condition does not hold, and the oblateness of the Earth actually causes major perturbations in the motion.

Outside a gravitating body the force function must satisfy Laplace's equation,

$$\nabla^2 U \equiv \partial^2 U/\partial x^2 + \partial^2 U/\partial y^2 + \partial^2 U/\partial z^2 = 0.$$

(This can be proved by differentiating equation (1.1.2); the summation is replaced by an integration.) It transpires that the force function of the body can normally be expanded in a power series in $1/r$, where r is the distance from its center of mass; the coefficients are called sperical harmonics. If, as is often the case, the body has symmetry about an axis, the force function can be expressed as

$$(1.1.3) \qquad \frac{Mk^2}{r}\left(1 - \frac{1}{r^2}J_2P_2 - \frac{1}{r^3}J_3P_3 - \cdots\right),$$

where the P_i are Legendre polynomials (functions of the latitude) and the J_i are constants; if the body is nearly spherical, the latter become small quite rapidly. Now it would be possible to find the force function of such a body by integration, if we knew precisely how it was put together. Failing this knowledge, it is still possible to write down its force function directly, so far as all the variable quantities are concerned. The theory of the motion of an artificial satellite, without drag, can be constructed using the force function (1.1.3). Then, later, observations may furnish the values of the J_i.

A lack of knowledge about the insides of a body is therefore no great hardship when its force function is required. (For more details, see [2, Chapter 4].)

1.2. *Newton's Laws of Motion.* We are concerned with Newtonian mechanics, the basic assumptions of which are contained in Newton's laws of motion. These are:

(1) Every particle continues in a state of rest or uniform motion in a straight line unless it is compelled by some external force to change that state.

(2) The rate of change of the linear momentum of a particle is proportional to the force applied to the particle and takes place in the same direction as that force.

(3) The mutual actions of any two bodies are always equal and oppositely directed.

A man who observes the motion of surrounding bodies that are not acted on by forces, and notes that they are not accelerated is entitled to feel that, for practical purposes, he is at rest with respect to some inertial system of reference. But if these bodies have any accelerations, then he is not (although he may invent forces such as centrifugal or Coriolis forces, to preserve the illusion). Certainly, no point fixed on the surface of the Earth could be the origin of an inertial system, although some sufficiently parochial experiments might give that impression. Motion observed in an ideal case by a nonrotating man at the center of the Earth would still show acceleration because of the action of the Sun, Moon, etc., on the Earth. Similarly, motion observed from the center of mass of the solar system should be affected by nearby stars, and the field of the galaxy (to say nothing about nearby galaxies): this is true in principle; but there is no known experiment to detect such effects, so that no purpose is served by considering acceleration with respect to the center of the galazy, and so on. So we shall not worry about the practical difficulties of choosing an inertial reference system, and we are certainly not concerned here with the thornier difficulties as to whether such a system can exist at all. We adopt the attitude that, given any problem in Newtonian mechanics, there exists an inertial system with respect to which the equations of motion can be written down; but no special assumption must be made about the whereabouts of the origin. Once the equations of motion have been set up, algebra will enable the origin to be transferred to this place or that. Also, inspection of some terms in the equations may result in their being rejected on account of their smallness. But the original equations must be written down without any assumptions being made about the origin, or the relative importance of different terms.

The measurement of "uniform motion" requires the use of a "uniformly flowing" time. The use of Universal Time (which is based on the rotation of the Earth) threw up accelerations of the Moon and planets that could not be explained by Newtonian mechanics, but which could result from nonuniform flowing of Universal Time. A suitable time has therefore been invented; this is Ephemeris Time. Its relation with Universal Time is given in the almanacs.

The second law can be applied only to motion observed with respect to an inertial reference system. If a particle of mass m is at $\mathbf{r}$ and the resultant of the forces acting on the particle is $\mathbf{F}$, then

$$(1.2.1) \qquad \mathbf{F} = \frac{d}{dt}\left(m \frac{d\mathbf{r}}{dt} \right).$$

Two important formulas follow from this. Firstly,

$$(1.2.2) \qquad \mathbf{r} \times \mathbf{F} = \frac{d}{dt}\left(\mathbf{r} \times m \frac{d\mathbf{r}}{dt} \right),$$

or "the moment of the external forces is equal to the rate of change of the angular momentum". Then, if $\mathbf{F}$ is the gradient of a force function U that does not contain the time explicitly, and if m is constant,

$$(1.2.3) \qquad \frac{1}{2}\left(\frac{d\mathbf{r}}{dt} \right)^2 - U = \text{constant}.$$

(For, differentiating (1.2.3) with respect to the time, we have the scalar product of $d\mathbf{r}/dt$ and (1.2.1).) This is the energy integral.

The third law is obeyed by Newton's law of gravitation, and is needed in a derivation of this law from Kepler's laws of planetary motion (quoted in §2.4).

Newton's laws apply directly to the motion of particles. If a body of finite extent is acted on by a system of forces, then the motion of its center of mass can be found by shifting the forces parallel to themselves so that their lines of action pass through the center of mass. The motion about the center of mass is considered, basically, through equation (1.2.2); subjects such as precession or physical libration fall under this heading; but they will not be considered here.

1.3. *Equations of Motion.* Consider the motion of n particles with fixed masses $m_1, m_2, \cdots, m_n$, which have position vectors $\rho_1, \rho_2, \cdots, \rho_n$, with respect to an inertial reference system. The equation of motion of m_i is

$$(1.3.1) \qquad m_i \rho_i'' = -k^2 m_i \sum_{j=1; j\neq i}^{n} m_j \frac{\rho_i - \rho_j}{\rho_{ij}^3},$$

where $\rho_{ij} = |\rho_i - \rho_j|$. A prime stands for differentiation with respect to the time. Adding the equations for all the particles, the forces cancel (from the algebra, or from Newton's third law) leaving

$$(1.3.2) \qquad \sum_{i=1}^{n} m_i \rho_i'' = 0.$$

But $\sum m_i \rho_i$ is proportional to the position vector of the center of mass of the system, and (1.3.2) says that this not accelerated with respect to the original inertial system; therefore the center of mass could be the origin of an inertial reference system.

Multiply (1.3.1) vectorially by $\mathbf{r}_i \times$, and add all n equations. The right-hand sides again cancel, leaving

$$(1.3.3) \qquad \sum_{i=1}^{n} m_i \mathbf{r}_i \times \mathbf{r}_i'' = 0, \quad \text{or} \quad \sum_{i=1}^{n} m_i \mathbf{r}_i \times \mathbf{r}_i' = \mathbf{h}$$

where $\mathbf{h}$ is a constant vector. The plane through the center of mass of the system and perpendicular to $\mathbf{h}$ is constant throughout the motion, and is called the "invariable plane" of the system.

The equations (1.3.1) can be written in the form

$$(1.3.4) \qquad m_i \rho_i'' = \nabla_i U,$$

where, if ρ_i has components (ξ_i, η_i, ζ_i), ∇_i has components $\partial/\partial \xi_i$, $\partial/\partial \eta_i$, $\partial/\partial \zeta_i$, and where

$$(1.3.5) \qquad U = k^2 \sum_{i<j=1}^{n} \sum \frac{m_i m_j}{\rho_{ij}}.$$

We therefore have the energy integral for the whole system,

(1.3.6) $$\frac{1}{2} \sum_{i=1}^{n} m_i \rho_i'^2 - U = \text{constant}.$$

But no integral can be written down for an individual member of the system.

Suppose that one mass, m_n, is considered to be dominant, either because of its relatively great magnitude, or because the motion in which we are interested takes place very close to it. Subtracting the equation of motion of m_n from that of m_i (after dividing by m_n and m_i, respectively) we find

$$\rho_i'' - \rho_n'' = -k^2 \sum_{j=1; j \neq i}^{n} m_j \frac{\rho_i - \rho_j}{\rho_{ij}^3} + k^2 \sum_{j=1}^{n-1} m_j \frac{\rho_n - \rho_j}{\rho_{nj}^3}$$

$$= -k^2 m_n \frac{\rho_i - \rho_n}{\rho_{in}^3} + k^2 m_i \frac{\rho_n - \rho_i}{\rho_{ni}^3}$$

$$- k^2 \sum_{j=1; j \neq i}^{n-1} m_j \left[\frac{\rho_i - \rho_j}{\rho_{ij}^3} - \frac{\rho_n - \rho_j}{\rho_{nj}^3} \right].$$

Now let the position vector of m_i with respect to m_n be $\mathbf{r}_i$, so that $\mathbf{r}_i = \rho_i - \rho_n$. Then

(1.3.7) $$\mathbf{r}_i'' + k^2(m_n + m_i) \frac{\mathbf{r}_i}{r_i^3} = k^2 \sum_{j=1; j \neq i}^{n-1} m_j \left[\frac{\mathbf{r}_j - \mathbf{r}_i}{r_{ij}^3} + \frac{\mathbf{r}_j}{r_j^3} \right].$$

Further, if

(1.3.8) $$R_{ij} = k^2 \left[\frac{1}{r_{ij}} - \frac{\mathbf{r}_i \cdot \mathbf{r}_j}{r_j^3} \right],$$

then

(1.3.9) $$\mathbf{r}_i'' + k^2(m_n + m_i) \frac{\mathbf{r}_i}{r_i^3} = \sum_{j=1; j \neq i}^{n-1} m_j \nabla_i R_{ij}.$$

Now if all the masses except m_n and m_i were zero, the right-hand sides of (1.3.7) or (1.3.9) would vanish, and the equations of motion would refer to the two-body problem; the solution of this is called Keplerian motion, and is described in the following chapter. It is frequently possible in celestial mechanics to find a dominant body, m_n, such that the terms on the right-hand side of (1.3.9) are much smaller than $k^2(m_n + m_i)\mathbf{r}_i/r_i^3$. In this case the motion can be considered as Keplerian motion "perturbed" by the forces on the right-hand side. This is why Keplerian motion is so important in celestial mechanics. The word "perturbation" normally implies a departure from Keplerian motion; the forces on the right-hand side of (1.3.9) are "perturbing forces" and the R_{ij} are "perturbative functions".

The reference system in (1.3.9) is noninertial. The terms on the right-hand side include the "direct" attractions of the bodies on m_i, and the "indirect" attractions on m_n, the origin. In a practical application many of these terms might be found to be negligible; but it can happen that the direct attraction of a body is negligible, but the indirect attraction is not. Further modifications can be made by adjusting the origin, and the mass of the dominant body; for details, see [2, §9.5].

2. The two-body problem.

2.1. *Properties of Conics.* Any orbit in the two-body problem is a conic, and before discussing the solution we shall briefly review the relevant properties of conics.

The polar equation of a conic can be written as

$$(2.1.1) \qquad\qquad p/r = 1 + e \cos f,$$

where the origin is at a focus of the conic, and f is the polar angle, measured from the major axis. [Equation (2.1.1) follows from the "focus-directrix" definition of a conic; i.e., that it is the locus of a point such that the ratio of its distance from a fixed point (a focus)

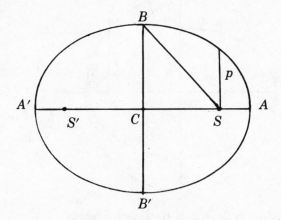

FIGURE 2. Features of an Ellipse

to its distance from a fixed line (a directrix) is constant, the value of the constant being equal to e.] e is the eccentricity; if $e = 0$ the conic is a circle; if e is less than one, it is an ellipse, which is bounded; if $e = 1$ it is a parabola; if e is greater than one it is a hyperbola.

Let an ellipse have center C, foci S, S', major axis AA', and minor axis BB', as shown in Figure 2. The following relations are useful, and should be memorized:

$$CA = CA' = a,$$
$$CB = CB' = b,$$
$$SA = q = a(1 - e),$$
$$SA' = q' = a(1 + e),$$
$$CS = CS' = ae,$$
$$p = a(1 - e^2),$$
$$b^2 = a^2(1 - e^2),$$
$$SB = a.$$

The ellipse can be obtained by the vertical projection of a circle. Let Q be a point on the circumference of the circle, and P the corresponding point on the ellipse, and let QP cut the major axis at R. Then $PR/QR = b/a$. Further, let $\angle ACQ = E$ (the "eccentric anom-

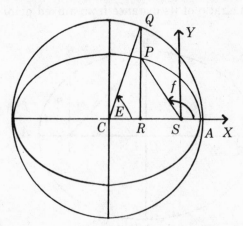

FIGURE 3. Orbital Reference System

aly"). With origin at S, let the X-axis point along SA, and the Y-axis point along the latus rectum, as shown in Figure 3. This reference system will be called the "orbital reference system". Then the coordinates of P can be written:

(2.1.2) $X = a(\cos E - e) = r \cos f, \qquad Y = b \sin E = r \sin f.$

The area of the ellipse is πab. We also have

(2.1.3) $r = \sqrt{(X^2 + Y^2)} = a(1 - e \cos E).$

Formulas for the parabola can be obtained from those for the ellipse by (carefully) letting $a \to \infty$ and $e \to 1$. It is safest first to eliminate a or e using $q = a(1 - e)$, since q remains finite. Suitable modifications to cover hyperbolic motion will be given in §2.4.

2.2. *The Solution of the Orbit.* Consider two particles of mass m_1 and m_2. Let the position vector of m_2 with respect to m_1 be $\mathbf{r}$. From (1.3.7) we see that the equation of motion of m_2 is

(2.2.1) $\mathbf{r}'' + k^2(m_1 + m_2)\mathbf{r}/r^3 = 0.$

If the origin were at the center of mass of the two bodies, the reference system (nonrotating) would be inertial. Then if the masses were at $\mathbf{r}_1$ and $\mathbf{r}_2$, the equation of motion of m_2 would be

$$m_2 \mathbf{r}_2'' = - k^2 m_1 m_2 \mathbf{r}/r^3$$

or, since $\mathbf{r}_2 = [m_1/(m_1 + m_2)]\, \mathbf{r}$,

(2.2.2) $\mathbf{r}_2'' = - k^2[m_1^3/(m_1 + m_2)^2]\,\mathbf{r}_2/r_2^3.$

Equations (2.2.1) and (2.2.2) are of the form

(2.2.3) $\mathbf{r}'' = - \mu \mathbf{r}/r^3$

but with different values of μ.

Equation (2.2.3) requires six constants of integration for its solution. Taking $\mathbf{r} \times$ (2.2.3), we find $\mathbf{r} \times \mathbf{r}'' = 0$, so that

(2.2.4) $\mathbf{r} \times \mathbf{r}' = \mathbf{h}$, a constant.

$\mathbf{h}$ supplies three arbitrary constants. From (2.2.4), $\mathbf{r} \cdot \mathbf{h} = 0$, which is the equation of a plane through the origin. The motion must take place in this plane; $\mathbf{h}$ determines its orientation, as well as the magnitude of the angular momentum. Now take $\mathbf{h} \times$ (2.2.3), and use

(2.2.4). We find

$$\mathbf{h} \times \mathbf{r}'' = -\frac{\mu}{r^3} (\mathbf{r} \times \mathbf{r}') \times \mathbf{r}$$

$$= -\frac{\mu}{r^3} \left[r^2 \mathbf{r}' - (\mathbf{r} \cdot \mathbf{r}') \mathbf{r} \right]$$

$$= -\frac{\mu}{r^3} \left[r^2 \mathbf{r}' - (rr') \mathbf{r} \right]$$

$$= -\mu \left[\mathbf{r}'/r - \mathbf{r} r'/r^2 \right]$$

$$= -\mu \frac{d}{dt} (\mathbf{r}/r)$$

$$= -\mu d\hat{r}/dt.$$

[$\mathbf{r}'$ is the velocity vector; r' is the rate of change of the scalar r. Differentiating $\mathbf{r}^2 = r^2$, we find $\mathbf{r} \cdot \mathbf{r}' = rr'$; a useful relation.] Integrating, we obtain

(2.2.4) $$\mathbf{h} \times \mathbf{r}' = -\mu \hat{r} - \mathbf{P},$$

where $\mathbf{P}$ is an arbitrary vector; but since it is perpendicular to $\mathbf{h}$, it only contains two arbitrary constants. The remaining constant of the motion will be considered in the following section. Taking $\mathbf{r} \cdot (2.2.4)$ we obtain

$$\mathbf{r} \cdot (\mathbf{h} \times \mathbf{r}') = -\mu r - \mathbf{P} \cdot \mathbf{r}$$

or

$$-\mathbf{h} \cdot (\mathbf{r} \times \mathbf{r}') = -\mu r - \mathbf{P} \cdot \mathbf{r}$$

or

$$h^2 = \mu r + \mathbf{P} \cdot \mathbf{r}$$

or

$$\frac{h^2/\mu}{r} = 1 + (\mathbf{P}/\mu) \cdot \hat{r}.$$

This is the same as equation (2.1.1). We have $h^2/\mu = p$, the vector $\mathbf{P}$ points along the major axis toward pericentron, and $P = \mu e$. The angle f is called the "true anomaly". If e is greater than one, only the branch of the hyperbola that is concave toward the origin can be described in the motion.

2.3. *The Orbit in Time.* The vector $\mathbf{r}'$ has components r' along $\hat{r}$ and rf' perpendicular to it; therefore the modulus of $\mathbf{r} \times \mathbf{r}'$ is $r^2 f'$, which is twice the rate of change of the area swept out by the radius vector. From (2.2.4) we have

$$(2.3.1) \qquad\qquad r^2(df/dt) = h.$$

The integration of this equation supplies the final constant of integration. Substituting for r from (2.1.1) we get a simple integral; but except when $e = 1$, it is convenient to introduce an intermediate angle, the eccentric anomaly.

Assume the motion to be elliptic. Differentiating (2.1.3) we find

$$r' = ae \sin E\, E'.$$

And differentiating $r \cos f = a(\cos E - e)$ (from (2.1.2)),

$$r' \cos f - r \sin f f' = -a \sin E\, E'.$$

Eliminating r' and f' from these two equations and (2.3.1) we find

$$h \sin f = a \sin E\, E' (1 + e \cos f)r.$$

Now using the relation

$$h^2 = \mu p = \mu a(1 - e^2),$$

and the formulas (2.1.2) and (2.1.3) to eliminate f and r, we find

$$\sqrt{(\mu/a^3)} = (1 - e \cos E)E',$$

which can be integrated to give

$$\sqrt{(\mu/a^3)}(t - T) = E - e \sin E,$$

where T is a constant of integration; it is equal to the time when $E = 0$, or when the body is at pericentron. This is Kepler's equation.

By letting the eccentric anomaly go from 0 to 2π, we get the time for a complete revolution, or the period of the motion, which is

$$(2.3.2) \qquad\qquad P = 2\pi \sqrt{(a^3/\mu)}.$$

The "mean motion", n, is defined by

$$(2.3.3) \qquad\qquad n = 2\pi/P,$$

so that $n^2 a^3 = \mu$. The angle

(2.3.4) $M = n(t - T)$

is defined as the "mean anamaly". So Kepler's equation can be written as

(2.3.5) $M = E - e \sin E$.

Normally we are given the time, and want to calculate E. That there is a unique solution can be seen from the fact that the right-hand side of (2.3.5) is monotonic increasing with E (for its differential coefficient with respect to E is $(1 - e \cos E)$, which is always positive). One of the best ways to find E is to use Newton's Method. If E_0 is a good guess, and E is correct, let

$$\Delta E = E - E_0,$$

and

$$\Delta M = M - M_0 = M - E_0 + e \sin E_0.$$

Then if $(\Delta E)^2$ is neglected,

$$\Delta E = \Delta M / (1 - e \cos E_0).$$

Because of the approximation, this correction is not exact, and the process will have to be repeated until ΔM becomes less than some small pre-assigned value. This process converges best when e is small, when a good first guess is

$$E_0 = M + e \sin M,$$

(although the series for E in terms of M and powers of e, given in the following chapter, can be truncated further along if desired). For more details, and for a discussion of the situation when e is nearly equal to one, see [3].

2.4. *Miscellaneous Properties.* Kepler's three laws of planetary motion are:

(1) The orbit of each planet is an ellipse, with the Sun at one of its foci. (Actually "Keplerian motion" is often now taken to include parabolic and hyperbolic motion, so that "conic" might replace "ellipse".)

(2) Each planet revolves so that the line joining it to the Sun sweeps out equal areas in equal intervals of time. (Therefore the acceleration of the planet is directed toward the Sun, and so also is

the force acting on the planet. From this law, and the first, Newton's law of gravitation can be deduced.)

(3) The squares of the periods of any two planets are in the same proportion as the cubes of their mean distances from the Sun. (This law should be modified so that $P^2(m_1 + m_2)/a^3$ is a constant for any two bodies, where a is the semimajor axis of the relative orbit, P is the period and m_1 and m_2 are the masses of the bodies. The law can be used to find the mass of a planet that has a satellite.)

Many important formulas for elliptic motion have been given already. A notable omission is the energy integral,

$$(2.4.1) \qquad\qquad \mathbf{r'}^2 = \mu(2/r - 1/a).$$

The parabolic velocity, or velocity of escape is found by putting $1/a = 0$. The circular velocity is found by putting $r = a$.

When changing from E to f or f to E, the following formulas are useful:

$$(2.4.2) \qquad \begin{aligned} \cos f &= (\cos E - e)/(1 - e \cos E), \\ \sin f &= \sqrt{(1 - e^2)}\sin E/(1 - e \cos E), \\ \cos E &= (e + \cos f)/(1 + e \cos f), \\ \sin E &= \sqrt{(1 - e^2)}\sin f/(1 + e \cos f). \end{aligned}$$

Using the relation $\tan^2(f/2) = (1 - \cos f)/(1 + \cos f)$, it is easy to verify that

$$(2.4.3) \qquad \tan(f/2) = \sqrt{\left(\frac{1+e}{1-e}\right)}\ \tan(E/2).$$

When using these formulas, it should be remembered that $f/2$ and $E/2$ always lie in the same quadrant. If we write

$$e = \sin\phi\ (0 \leqq \phi < \pi/2),$$

as is commonly done, then

$$\tan(f/2) = \tan(\pi/4 + \phi/2)\tan(E/2).$$

From Kepler's equation, and (2.1.3) we have

$$(2.4.4) \qquad\qquad E' = na/r.$$

Also we have

(2.4.5) $$r' = na^2 e \sin E/r = (e\mu/h)\sin f.$$

Formulas (2.1.2) are important. Differentiating them, we find

(2.4.6) $X' = -na^2 \sin E/r,$ $Y' = na^2 \sqrt{(1 - e^2)} \cos E/r.$

In parabolic motion let q be the pericentron distance, then the equation of the orbit is

$$r = q \sec^2(f/2).$$

An equation involving the time is

$$\frac{1}{3} \tan^3(f/2) + \tan(f/2) = \sqrt{(\mu/2q^3)}(t - T).$$

Formulas for hyperbolic motion can be derived from those for elliptic motion as follows. Assume a to be negative for a hyperbola. If $i^2 = -1$, replace E by iF (so that $\cos E$ becomes $\cosh F$ and $\sin E$ becomes $i \sinh F$), replace n by $-i\nu$, where $\nu^2 a^3 = -\mu$, and ν is positive, and replace $\sqrt{(1 - e^2)}$ by $i\sqrt{(e^2 - 1)}$.

Many important formulas have been omitted here. The reader should consult, in particular, [3].

2.5. *The Orbit in Space.* An orbit is defined by six constants, and these require some kind of reference system. The celestial equator or ecliptic are often used as reference planes, with the direction of the vernal equinox defining an axis. Neither of these planes is fixed, and it is necessary to use their mean positions for some definite epoch.

A suitable set of constants would be the components of position and velocity, r_0, r'_0, at some time t_0; it is possible to calculate from these the position r at any time t (formulas for the calculation of the velocity are easily deduced and will not be given here). Since the motion takes place in a plane, it must be possible to resolve r along the directions of r_0 and r'_0. So we can write

(2.5.1) $$\mathbf{r} = f\mathbf{r}_0 + g\mathbf{r}'_0,$$

where f and g are scalar functions of t_0 and t and the initial conditions. From (2.5.1) we find

$$f\mathbf{h} = \mathbf{r} \times \mathbf{r}'_0 \quad \text{and} \quad g\mathbf{h} = \mathbf{r}_0 \times \mathbf{r}.$$

These are vector equations, independent of the reference system.

So f and g can be evaluated by substituting components referred to the "orbital reference system" defined in §2.1. The components are given by formulas (2.1.2) and (2.4.6). After substitution and some simplification, we find

(2.5.2)

$$f = \frac{a}{r}[\cos(E - E_0) - e \cos E_0],$$

$$g = \frac{1}{n}[\sin(E - E_0) - e(\sin E - \sin E_0)].$$

Before using (2.5.1) and (2.5.2) to calculate $\mathbf{r}$, it would be necessary to calculate a, e, E_0, and E. We are given $\mathbf{r}_0$ and $\mathbf{r}_0'$. (2.4.1) will give a. Then (2.1.3) and (2.4.5) will give $e \cos E_0$ and $e \sin E_0$, from which e and E_0 can be found. (In using (2.4.5), remember that $r_0 r_0' = \mathbf{r}_0 \cdot \mathbf{r}_0'$.) Finally, Kepler's equation can be used to find E.

In the formulas above a, e, and E_0 are introduced as intermediate elements; but they help to give a picture of the shape and size of the orbit, and the initial whereabouts in the orbit, that $\mathbf{r}_0$ and $\mathbf{r}_0'$ com-

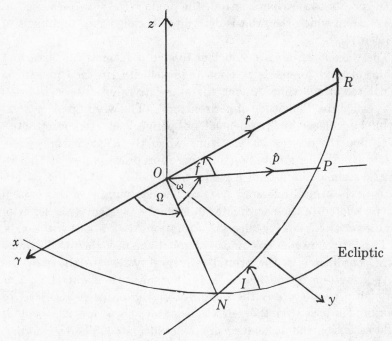

FIGURE 4. Geometrical Elements.

pletely fail to do. It is more usual to use six constants, each of which has an easily visualized geometrical meaning; these are the "geometrical elements" of the orbit. a and e are two possible elements, and a third is a time of pericentron passage, T, or any number, such as the mean anomaly at some time, that enables the position in the ellipse to be found at any time. The description of the orientation of the orbit in space requires three angles, illustrated in Figure 4. In the figure the fundamental plane is the ecliptic (it could equally well be the celestial equator, if preferred), the Sun is at O, Ox points toward the vernal equinox and Oz toward the north pole of the ecliptic. The plane of the orbit cuts the celestial sphere in the great circle NPR where N is the point where the body in its orbit crosses the ecliptic, going north; it is called the "ascending node". The angle xON (measured eastward around the ecliptic) is called the "longitude of the ascending node" and is written as Ω. The angle between the ecliptic and the plane of the orbit is the "inclination", I. For $0 \leq I < 90°$ the orbit is direct; for $90° < I < 180°$, it is retrograde. If OP points toward pericentron, the angle $NOP = \omega$ (measured in the sense in which the orbit is described) is called the "argument of pericentron".

These six constants are sufficient to give a geometrical picture of the orbit, and to enable position (and velocity) in the orbit to be calculated at any time. Among the alternatives often used is $\varpi = \Omega + \omega$, called the "longitude of pericentron". (The word "pericentron" would be replaced by "perihelion", or "perigee", etc. as appropriate.)

To find the position at any time when the elements are given, first solve Kepler's equation for the appropriate value of the eccentric anomaly, and then use equations (2.1.2) to find the coordinates in the orbital reference system. The coordinates in this system can be related to the coordinates in any other system by a series of rotations. The following successive rotations: $-\omega$ about the Z-axis, $-I$ about the new x-axis, and $-\Omega$ about the new z-axis, will transform coordinates in the orbital reference system to those in the x-, y-, z-system of the figure. A further rotation about the x-axis through $-\epsilon$ (where ϵ is the obliquity of the ecliptic) will lead to coordinates based on the celestial equator; these are necessary if right ascension and declination are to be calculated. The transformation resulting from a rotation about an axis of reference can be most conveniently described by a matrix multiplication. For details, see

[2, Appendix B]. The result of all the rotations described above can be written in the form

(2.5.3)
$$\begin{bmatrix} x \\ y \\ z \end{bmatrix} = \begin{bmatrix} P_x & Q_x \\ P_y & Q_y \\ P_z & Q_z \end{bmatrix} \begin{bmatrix} X \\ Y \end{bmatrix},$$

where the P's and Q's are direction cosines of the X- and Y-axes with respect to the x-, y-, z-axes.

Suppose that it is required to find the geometrical elements when position and velocity, $\mathbf{r}_0$, $\mathbf{r}_0'$, are given for a time t_0. a is found from (2.4.1) and e and E_0 from (2.1.3) and (2.4.5), as before. Then T, or M_0, the mean anomaly at the epoch, can be found from Kepler's equation. The individual angles Ω, ω, and I might now be found from the formulas:

(2.5.4)
$$\mathbf{h} = \mathbf{r}_0 \times \mathbf{r}_0'$$
$$= (h_x, h_y, h_z),$$
$$h_x = h \sin \Omega \sin I,$$
$$h_y = - h \cos \Omega \sin I,$$
$$h_z = h \cos I.$$

f from (2.4.3),
$$\sin u = \frac{z}{r} \operatorname{cosec} I,$$
$$r \cos u = x \cos \Omega + y \sin \Omega,$$

(u is the "argument of the latitude")
$$\omega = u - f,$$

where an extra 360° may have to be added to make ω lie between 0 and 360°.

Alternatively, it may be better to find the P's and Q's of (2.5.3) directly. (2.5.3) can be written more generally as

$$\begin{bmatrix} x_0 & x_0' \\ y_0 & y_0' \\ z_0 & z_0' \end{bmatrix} = \begin{bmatrix} P_x & Q_x \\ P_y & Q_y \\ P_z & Q_z \end{bmatrix} \begin{bmatrix} X_0 & X_0' \\ Y_0 & Y_0' \end{bmatrix},$$

where $X_0 = a(\cos E_0 - e)$, etc. from (2.1.2) and (2.4.6). Then, solving for the P's and Q's, we find (since $X_0 Y_0' - Y_0 X_0' = h$),

(2.5.5)
$$\begin{bmatrix} P_x & Q_x \\ P_y & Q_y \\ P_z & Q_z \end{bmatrix} = \begin{bmatrix} x_0 & x_0' \\ y_0 & y_0' \\ z_0 & z_0' \end{bmatrix} \begin{bmatrix} Y_0'/h & -X_0'/h \\ -Y_0/h & X_0/h \end{bmatrix}.$$

The individual angles Ω, ω, and I can also be determined from the P's and Q's.

For an account of the determination of the elements when two positions for two different times are given, see [3].

In certain cases some element can only be poorly determined. For instance, if e is small, E_0 and ω or ϖ cannot be found as accurately as the other elements because, somewhere along the line, their calculation involves division by e. Similarly, if I is small, Ω is poorly determined. It is possible to put too much emphasis on the difficulties that result. ω or Ω should not be considered as goals in themselves. Suppose that the object of the work is to calculate position and velocity at any time; then it need not matter that for small e an angle such as ω is poorly determined (in fact there will be a multiplication by e during the calculation), and the accuracy of the final result need not suffer at all. Difficulties due to a small I can be avoided by using the P's and Q's. If a programmer is determined to avoid any division by e, there are several ways in which this can be achieved. One possibility is to use $e \cos E_0$ and $e \sin E_0$ as elements; there need be no doubt about their accuracy. Let E be the eccentric anomaly at time t, then from Kepler's equation applied to the times t_0 and t, we find

$$n(t - t_0) = E - E_0 - e \sin E + e \sin E_0$$

(2.5.6)
$$= \Delta E - e \cos E_0 \sin \Delta E - e \sin E_0 \cos \Delta E + e \sin E_0,$$

where $\Delta E = E - E_0$. This can be solved for ΔE, and then $e \sin E$ and $e \cos E$ can be calculated, and (2.5.2) and (2.5.1) used to find the position at time t.

If the elements are to be considered as slowly varying quantities in perturbed motion, other problems may arise, and different elements are needed for special cases.

3. **Expansions in series.**

3.1. *Expansions in Powers of the Eccentricity.* The stumbling block in any attempt to express position in Keplerian motion explicitly in terms of the time comes in any attempt to express the eccentric anomaly explicitly as a function of the mean anomaly. In general it cannot be done in a finite number of terms. But if the eccentricity is sufficiently small, approximate expressions can be developed that are good enough. Fortunately, nearly all the planets and satellites in the solar system have orbits with moderately small eccentricities.

For a circular orbit, $E = M$. If e is small, then, writing Kepler's equation in the form

$$E = M - e \sin E,$$

we see that to the order of e, we can put

$$E_1 = M + e \sin M.$$

Now if we put $E_2 = E_1 + \delta E_1$, and ignore e^3, we find

$$E_2 = M + e \sin M + \frac{1}{2} e^2 \sin 2M.$$

Further development along these lines becomes immensely tedious, and it would be an advantage if some formula could be found that would give the general term. Such a formula is given by Lagrange's theorem, which can be stated for the problem in hand as follows: Let

$$E = M + ef(E),$$

then

$$F(E) = F(M) + \frac{e}{1!} f(M) F'(M) + \frac{e^2}{2!} \frac{d}{dM} \{ [f(M)]^2 F'(M) \} + \cdots$$

$$+ \frac{e^q}{q!} \frac{d^{q-1}}{dM^{q-1}} \{ [f(M)]^q F'(M) \} + \cdots.$$

Now put $F(E) \equiv E$, so that $F'(E) = dF/dE = 1$; and put $f(E) \equiv \sin E$. Then we get

$$E = M + \frac{e}{1!} \sin M + \frac{e^2}{2!} \frac{d}{dM} (\sin^2 M) + \cdots$$

(3.1.1)
$$+ \frac{e^n}{n!} \frac{d^{n-1}}{dM^{n-1}} (\sin^{n+1} M) + \cdots.$$

Any other $F(E)$, such as $r \equiv F(E) = a(1 - e \cos E)$, can be expanded similarly.

The general statement of Lagrange's theorem would be useful, because it includes the condition for convergence of the series (and it is not often that a question of the convergence of a series in celestial mechanics can be answered). Limitation of space prevents a discussion here, but see [4, §46]. The upshot is that series in powers of the eccentricity converge for values of e less than $0.6627\cdots$.

3.2. *Applications of Lagrange's Theorem.* An unattractive feature of (3.1.1) is that powers of trigonometric functions appear. It is usually simpler to deal with terms such as $\sin kM$ rather than $\sin^k M$, so that Fourier series are generally preferable to power series. Also, it is laborious to change from one to the other, so that it is an advantage if a Fourier series can be generated in the first place. One way of doing this is to use the exponential function. For

$$E^{ikM} = \cos kM + i \sin kM,$$

where E is the exponential and $i^2 = -1$, so that what is generated as a power series in E^{iM} becomes a Fourier series.

Consider (2.4.3). It is usual in these developments to get rid of the square root, so we introduce

(3.2.1) $$\frac{1+\beta}{1-\beta} = \sqrt{\left(\frac{1+e}{1-e}\right)}$$

so that

(3.2.2) $$\beta = \frac{1}{2} e(1 + \beta^2).$$

Also, introducing $\sin\phi = e$, we have $\beta = \tan\frac{1}{2}\phi$. (3.2.2) could equally well have been written as

$$\beta = M + \left(\frac{1}{2}e\right)(1 + \beta^2), \qquad M = 0.$$

Then β^j can be expanded in powers of $\frac{1}{2}e$ by Lagrange's theorem to give

$$\beta^j = \sum_{q=1}^{\infty} \left[\left(\frac{1}{2}e\right)^q \middle/ q'\right]\left\{\frac{d^{q-1}}{dM^{q-1}}\left[(1+M^2)^q jM^{j-1}\right]\right\}_{M=0}$$

$$= j\sum_{q=1}^{\infty} \left[\left(\frac{1}{2}e\right)^q \middle/ q'\right]\left\{\frac{d^{q-1}}{dM^{q-1}}\left[\sum_{p=0}^{q}\frac{q!}{(q-p)!p!}M^{2p+j-1}\right]\right\}_{M=0}.$$

For a term to survive the operation $M = 0$ after the differentiation, we must have $2p + j - 1 = q - 1$. Then for a definite value of p, $q = 2p + j$; so we can write

$$\beta^j = j \sum_{p=0}^{\infty} \left(\frac{1}{2} e\right)^{2p+j} \frac{(2p + j - 1)!}{(p + j)!\, p!}$$

(3.2.3)

$$= \left(\frac{1}{2} e\right)^j \left[1 + \left(\frac{1}{2} e\right)^2 j + \left(\frac{1}{2} e\right)^4 \frac{j(j + 3)}{2!} + \cdots\right].$$

We are now in a position to consider expansions in powers of β.
  For the applications, put

(3.2.4) $if = \log x, \qquad iE = \log y, \qquad iM = \log z,$

where $i^2 = -1$, and the logs are to the base E, so that $x = E^{if}$, etc. Then

$$x^k + 1/x^k = 2 \cos kf, \qquad x^k - 1/x^k = 2i \sin kf, \text{ etc.}$$

  From (2.4.3) and (3.2.1) we have

$$\frac{x - 1}{x + 1} = \frac{1 + \beta}{1 - \beta} \frac{y - 1}{y + 1},$$

so

(3.2.5) $x = \dfrac{y - \beta}{1 - \beta y}, \quad \text{or} \quad y = \dfrac{x + \beta}{1 + \beta x}.$

Then from the first of these,

$$\log x = \log y + \log(1 - \beta/y) - \log(1 - \beta y),$$

and, bearing in mind that for $|z| < 1$, $\log(1 + z) = z - z^2/2 + z^3/3 + \cdots$, we can write this as

$$\log x = \log y + \beta(y - 1/y) + \frac{1}{2}\beta^2(y^2 - 1/y^2) + \cdots,$$

so that, from (3.2.4) we have (after division by i)

(3.2.6) $f = E + 2\left(\beta \sin E + \dfrac{1}{2}\beta^2 \sin 2E + \dfrac{1}{3}\beta^3 \sin 3E + \cdots\right).$

From (3.2.5) we see that to exchange x and y it is sufficient to change the sign of β. Therefore

(3.2.7) $E = f - 2\left(\beta \sin f - \dfrac{1}{2}\beta^2 \sin 2f + \dfrac{1}{3}\beta^3 \sin 3f + \cdots\right).$

Substituting from (3.2.4) into Kepler's equation, we have

$$\log z = \log y - \frac{1}{2} e(y - 1/y).$$

Eliminating y, from (3.2.5), and using (3.2.2) to eliminate e, we can transform this to

$$\log z = \log x + \log(1 + \beta/x) - \log(1 + \beta x) - \frac{\beta}{1 + \beta^2} \frac{(1 - \beta^2)(x^2 - 1)}{(x + \beta)(1 + \beta x)}.$$

Now $(1 - \beta^2)/(1 + \beta^2) = \sqrt{(1 - e^2)} = \cos\phi$, so that the final term on the right-hand side can be written as

$$- \beta \cos \phi [\, x/(1 + \beta x) - (1/x)/(1 + \beta/x) \,]$$

$$= \beta \cos\phi \left[\frac{1}{x}(1 - \beta/x + \beta^2/x^2 - \cdots) - x(1 - \beta x + \beta^2 x^2 - \cdots) \right].$$

Therefore, expanding the logarithms as before, and substituting from (3.2.4), we get

$$M = f - 2\left(\beta \sin f - \frac{1}{2}\beta^2 \sin 2f + \cdots \right)$$

$$+ 2\beta \cos\phi\, (-\sin f + \beta \sin 2f - \cdots)$$

(3.2.8)

$$= f - 2\left[\beta(1 + \cos\phi) \sin f - \beta^2 \left(\frac{1}{2} + \cos\phi \right) \sin 2f \right.$$

$$\left. + \beta^3 \left(\frac{1}{3} + \cos\phi \right) \sin 3f + \cdots \right].$$

The difference between the mean and true anomalies is called the "equation of the center".

3.3. *Fourier Series.* The derivation of the series in the preceding section was a trifle roundabout. Before proceeding with the direct derivation of Fourier series we shall briefly state enough theorems to build up the relevant background.

Let $f(t)$ be a periodic function with bounded variation and period 2π; let it be integrable for all t, so that the products $f(t) \sin pt$, $f(t) \cos pt$ are also integrable. Define

$$a_0 = \frac{1}{2\pi} \int_0^{2\pi} f(t)dt,$$

and

$$a_p = \frac{1}{\pi} \int_0^{2\pi} f(t) \cos pt \, dt, \qquad b_p = \frac{1}{\pi} \int_0^{2\pi} f(t) \sin pt \, dt.$$

The series

(3.3.1) $$a_0 + \sum_{p=1}^{\infty} (a_p \cos pt + b_p \sin pt)$$

is called the Fourier series of $f(t)$. If $f(t)$ is continuous, its sum is equal to $f(t)$. Furthermore, if its derivative is bounded, then the Fourier series is uniformly convergent.

If the form (3.3.1) is accepted, then the formula for the coefficients is very easily recovered by multiplying through by $\cos pt$ or $\sin pt$ and integrating from 0 to 2π, so that every term but one vanishes. If $f(t)$ is an even function, then only the a_p appear, and it is sufficient to integrate from 0 to π, and divide by $\pi/2$. Similarly, if $f(t)$ is an odd function, only the b_p appear.

Using the exponential function, we could also put

(3.3.2) $$f(t) = \sum_{p=-\infty}^{+\infty} \alpha_p E^{ipt},$$

where

$$\alpha_p = \frac{1}{2\pi} \int_0^{2\pi} E^{-ipt} f(t) \, dt.$$

(Note that as soon as trigonometric functions are replaced by exponential functions, the summations must go from minus infinity to plus infinity.)

Consider the expansion of the function a/r as a Fourier series in the mean anomaly. It is an even function of E, and consequently of M. Also $a/r = dE/dM$. Therefore

$$\frac{a}{r} = \frac{1}{\pi} \int_0^{\pi} \frac{a}{r} \, dM + \frac{2}{\pi} \sum_{p=1}^{\infty} \cos pM \int_0^{\pi} \frac{a}{r} \cos pM \, dM$$

$$= \frac{1}{\pi} \int_0^{\pi} dE + \frac{2}{\pi} \sum_{p=1}^{\infty} \cos pM \int_0^{\pi} \cos(pE - pe \sin E) dE.$$

Define the "Bessel's coefficient" $J_p(x)$, of order p and argument x by

(3.3.3) $$J_p(x) = \frac{1}{\pi} \int_0^\pi \cos(p\phi - x \sin \phi)d\phi.$$

Then we can write

(3.3.4) $$\frac{a}{r} = 1 + 2 \sum_{p=1}^\infty J_p(pe) \cos pM.$$

These coefficients are ubiquitous, and it is necessary to break off and derive some of their properties before continuing to develop any other series.

3.4. *Properties of Bessel's Functions.* $J_p(x)$ was defined in (3.3.3). But since

$$\frac{1}{2\pi} \int_0^{2\pi} \sin(p\phi - x \sin \phi)d\phi = 0,$$

we could have written

(3.4.1) $$J_p(x) = \frac{1}{2\pi} \int_0^{2\pi} E^{-ip\phi} E^{ix \sin \phi} d\phi.$$

Now suppose that we wanted to expand the function $E^{ix \sin \phi}$ as

$$E^{ix \sin \phi} = \sum_{p=-\infty}^{+\infty} \alpha_p E^{ip\phi}.$$

Then, from (3.3.2) we would find that $\alpha_p = J_p$, so that

(3.4.2) $$E^{ix \sin \phi} = \sum_{p=-\infty}^{+\infty} J_p(x) E^{ip\phi},$$

a formula that can be useful, incidentally, where trigonometric functions of trigonometric functions are concerned.

Now put $E^{i\phi} = z$, so that $2i \sin \phi = z - 1/z$. Then (3.4.2) becomes

(3.4.3) $$\exp \left[\frac{x}{2} (z - 1/z) \right] = \sum_{p=-\infty}^{+\infty} J_p(x) z^p.$$

The left-hand side of (3.4.3) can be written as the product of

$$\sum_{\alpha=0}^{\infty} \left(\frac{1}{2}x\right)^{\alpha} z^{\alpha}/\alpha! \quad \text{and} \quad \sum_{\beta=0}^{\infty} (-1)^{\beta}\left(\frac{1}{2}x\right)^{\beta} z^{-\beta}/\beta!.$$

To find the coefficient of z^p put $\alpha = \beta + p$. Now α cannot be negative, so that for $p \geqq 0$,

$$(3.4.4) \qquad J_p(x) = \sum_{\beta=0}^{\infty} (-1)^{\beta} \frac{1}{\beta!(\beta+p)!} \left(\frac{1}{2}x\right)^{p+2\beta}.$$

For $p < 0$, the summation runs from $\beta = -p, \; -p+1, \cdots$. The series (3.4.4) is absolutely convergent for all x.

In (3.4.3) change z to $-z$, and x to $-x$; the left-hand side is the same, so that

$$J_p(x) = (-1)^p J_p(-x).$$

Also, change z to $-1/z$. The left-hand side is still the same, so that

$$J_p(x) = (-1)^p J_{-p}(x).$$

Combining these two results, we find

$$(3.4.5) \qquad J_{-p}(-x) = J_p(x).$$

Differentiating (3.4.3) with respect to z, and using (3.4.3) to remove the exponential on the left-hand side, we get

$$\frac{1}{2}x(1+1/z^2) \sum_{p=-\infty}^{\infty} J_p(x) z^p = \sum_{p=-\infty}^{\infty} p J_p(x) z^{p-1}.$$

So that from the coefficients of z^{p-1}, we find

$$(3.4.6) \qquad \frac{1}{2}x[J_{p-1}(x) + J_{p+1}(x)] = p J_p(x).$$

Similarly, differentiating (3.4.3) with respect to x, and considering the coefficients of z^p, we can find

$$(3.4.7) \qquad \frac{1}{2}[J_{p-1}(x) - J_{p+1}(x)] = J_p'(x).$$

Differentiating (3.4.7) with respect to x, we have

$$J_p''(x) = \frac{1}{2}[J_{p-1}'(x) - J_{p+1}'(x)]$$

$$= \frac{1}{4}[J_{p-2}(x) - 2J_p(x) + J_{p+2}(x)] \qquad \text{(from (3.4.7))}$$

$$= \frac{1}{4}\left[\frac{2}{x}(p-1)J_{p-1}(x) - J_p(x) - 2J_p(x) + \right.$$

$$\left. + \frac{2}{x}(p+1)J_{p+1}(x) - J_p(x)\right] \qquad \text{(from (3.4.6))}$$

$$= -J_p(x) + \frac{1}{2x}[(p-1)J_{p-1}(x) + (p+1)J_{p+1}(x)]$$

$$= -J_p(x) + \frac{p^2}{x^2}J_p(x) - \frac{1}{x}J_p'(x) \qquad \text{(from (3.4.7))}.$$

So J_p is a solution of the equation

$$y'' + \frac{1}{x}y' + (1 - p^2/x^2)y = 0.$$

The general theory of Bessel's functions can start from this equation; but this is not needed for our purpose. We need only the solutions of the first kind, with integral values of p, and the definition given above is sufficient.

The series (3.4.4) demonstrates that the J_p can always be calculated. But there are many alternative methods of calculation, using such devices as recurrence relations, or continued fractions. See [1].

3.5. *Applications of Bessel's Functions.* Consider the expansion of $\sin mE$. It is an odd function of E or M, so that

$$\sin mE = \frac{2}{\pi}\sum_{p=1}^{\infty}\sin pM\int_0^{\pi}\sin mE\,\sin pM\,dM.$$

Now $\sin pM\,dM = -(1/p)d(\cos pM)$, so that, introducing E, we can write

$$\sin mE = -\frac{2}{\pi}\sum_{p=1}^{\infty}\frac{\sin pM}{p}\int_0^{\pi}\sin mE\,d[\cos(pE - pe\sin E)],$$

and, integrating by parts,

$$\sin mE = -\frac{2}{\pi} \sum_{p=1}^{\infty} \frac{\sin pM}{p} \left\{ \left[\sin mE \cos(pE - pe \sin E) \right]_0^\pi \right.$$

$$\left. - \int_0^\pi m \cos mE \cos(pE - pe \sin E) \, dE \right\}.$$

The integrated term vanishes at the limits; using the formula for the product of two cosines, the integrand can be developed to give

$$\sin mE = \frac{m}{\pi} \sum_{p=1}^{\infty} \frac{\sin pM}{p} \int_0^\pi \left\{ \cos\left[(p + m)E - pe \sin E \right] \right.$$

$$\left. + \cos\left[(p - m)E - pe \sin E \right] \right\} dE$$

$$= m \sum_{p=1}^{\infty} \frac{\sin pM}{p} \left\{ J_{p-m}(pe) + J_{p+m}(pe) \right\}.$$

When $m = 1$, we have, by (3.4.6),

(3.5.1)
$$\sin E = \frac{2}{e} \sum_{p=1}^{\infty} \frac{\sin pM}{p} J_p(pe).$$

Similarly, we find

$$\cos mE = a_0 + \frac{2}{\pi} \sum_{p=1}^{\infty} \cos pM \int_0^\pi \cos mE \cos pM \, dM$$

$$= a_0 + \frac{2}{\pi} \sum_{p=1}^{\infty} \cos pM \int_0^\pi \frac{m}{p} \sin mE \sin(pE - pe \sin E) \, dE$$

(after integration by parts)

$$= a_0 + m \sum_{p=1}^{\infty} \frac{\cos pM}{p} \left\{ J_{p-m}(pe) - J_{p+m}(pe) \right\}.$$

Here

$$a_0 = \frac{1}{\pi} \int_0^\pi \cos mE \, dM = \frac{1}{\pi} \int_0^\pi \cos mE \, (1 - e \cos E) \, dE$$

$$= \frac{1}{\pi} \int_0^\pi \left[\cos mE - \frac{1}{2} e \cos(m + 1)E - \frac{1}{2} e \cos(m - 1)E \right] dE$$

$$= 1 \text{ if } m = 0; \; -e/2 \text{ if } m = 1; \; 0 \text{ if } m > 1.$$

In particular, using (3.4.7),

$$(3.5.2) \qquad \cos E = -\frac{1}{2} e + 2 \sum \frac{\cos pM}{p} J'_p(pe).$$

We now have enough formulas to expand quite a lot of functions as Fourier series in the mean anomaly. For instance, Kepler's equation combined with (3.5.1) will cope with E. r/a can be expanded using (2.1.3) and (3.5.2). X and Y of (2.1.2) can be found similarly. Sometimes a little ingenuity can help; in seeing, for instance, that $X/r^3 = -a^{-3} d^2 X/dM^2$, $Y/r^3 = -a^{-3} d^2 Y/dM^2$. Another example is

$$\sin f = \sqrt{(1 - e^2)} \sin E / (1 - e \cos E) = \cot \phi \, \frac{d}{dM}\left(\frac{r}{a}\right).$$

$\cos f$ is easily found from (2.1.1) and (3.3.4). A function such as $(r/a)^2$ can be easily written down in terms of a Fourier series in E, and from there to one in M. And so on. Many more examples are given in [1] and [4].

It should be noted that these Fourier series are valid for any value of the eccentricity; but if they are re-arranged as power series in the eccentricity, then the upper limit noted in §3.1 applies.

In the series for a/r or r/a or powers of these, it is noticeable that the lowest power of e in any coefficient is equal to the multiple of M in that term; this fact is a great help when deciding where to truncate a series. Although the equality just pointed out is not general, the fact that the lowest order of e increases as the coefficient of M is; this is a characteristic of these expansions stressed by D'Alembert, which now bears his name. In the expansion of $(a/r)^{\pm k}$ times the sine or cosine of m times the eccentric or true anomaly, the lowest power of e is in general equal to the coefficient of M minus m.

3.6. *Postscript.* The reader should be warned that the above notes are extremely incomplete. No mention has been made of expansions in powers of the time; nor of the first-order differences between two "nearly equal" elliptic orbits. These fall usually under the heading of "orbit determination" and are dealt with more than adequately in [3]. Nothing has been said about the proper choice of units, even though, without this, an attempt at practical calculation in celestial mechanics may be stillborn. Also hyperbolic orbits have been neglected, in spite of their increasing importance. Bearing in mind these and other omissions, the reader should consult some of the references.

References

1. D. Brouwer and G. M. Clemence, *Methods of celestial mechanics,* Academic Press, New York, 1961.

2. J. M. A. Danby, *Fundamentals of celestial mechanics,* Macmillan, New York, 1962.

3. P. Herget, *The computation of orbits,* Univ. of Cincinnati, Cincinnati, Ohio, 1948.

4. H. C. Plummer, *Dynamical astronomy,* Dover, New York, 1960. (First published, 1918.)

5. G. M. Clemence, *Planetary distances according to general relativity,* Astronom. J. **67** (1962), 379.

YALE UNIVERSITY OBSERVATORY

J. M. A. Danby

Matrix Methods

Consider the system of n first order differential equations

(1) $dX_i/dt = f_i(X_1, X_2, \cdots, X_n; t), \quad i = 1, 2, \cdots, n,$

relating the n coordinates X_i and the time t. These can be written symbolically in the condensed form

(1A) $\mathbf{X}' = \mathbf{f}(\mathbf{X}, t),$

where $\mathbf{X}$ and $\mathbf{f}$ are column matrices; the prime represents differentiation with respect to the time.

Suppose that a solution $\mathbf{X}_R(t)$ has been found, having initial conditions

$$\mathbf{X}_R(t_0) = \mathbf{X}_0.$$

A "slightly different" solution, $\mathbf{X}_R + \delta\mathbf{X}$, might be found by solving equations (1) again, subject to initial conditions $\mathbf{X}_0 + \delta\mathbf{X}_0$ at t_0. Then δX would be found by subtracting $\mathbf{X}_R$. But this approach can be extravagant in significant figures, and it is often better to solve directly for $\delta\mathbf{X}$.

If the squares and products of small quantities are neglected, then $\delta\mathbf{X}$ must satisfy the first variational equations of the system (1):

32

$$(2) \qquad \begin{bmatrix} \delta X_1' \\ \delta X_2' \\ \cdot \\ \cdot \\ \cdot \\ \delta X_n' \end{bmatrix} = \begin{bmatrix} \partial f_1/\partial X_1 & \partial f_1/\partial X_2 & \cdots & \partial f_1/\partial X_n \\ \partial f_2/\partial X_1 & \partial f_2/\partial X_2 & \cdots & \partial f_2/\partial X_n \\ \cdot & \cdot & \cdots & \cdot \\ \cdot & \cdot & \cdots & \cdot \\ \cdot & \cdot & \cdots & \cdot \\ \partial f_n/\partial X_1 & \partial f_n/\partial X_2 & \cdots & \partial f_n/\partial X_n \end{bmatrix} \begin{bmatrix} \delta X_1 \\ \delta X_1 \\ \cdot \\ \cdot \\ \cdot \\ \delta X_n \end{bmatrix}$$

or

$$(2A) \qquad \delta \mathbf{X}' = A\,\delta \mathbf{X}.$$

The solution $\mathbf{X}_R(t)$ will be called the "reference orbit". Each of the partial differential coefficients in the n-by-n matrix A is evaluated along the reference orbit, so that A is a known function of the time.

Equations (2) are solved when any set of n linearly independent solutions is known. Finding these may present difficulties; but suppose for the moment that we have such a set, and that it consists of the separate columns of the matrix with elements $\delta x_{ij}(t)$. Since any linear combination of these columns also gives a solution, the columns of

$$(3) \qquad \Omega(t_0, t) \equiv \left[\delta x_{ij}(t)\right]\left[\delta x_{ij}(t_0)\right]^{-1}$$

must all be solutions. The matrix $\Omega(t_0, t)$ is equal to the identity matrix when $t = t_0$; this provides the necessary initial conditions to find its components by numerical integration. For example, equations (2) would be solved subject to the initial conditions $\delta X_1 = 1$, $\delta X_i = 0$ $(i \neq 1)$ to give the first column. The initial conditions for the second column would be $\delta X_2 = 1$, $\delta X_i = 0$ $(i \neq 2)$; and so on. $\Omega(t_0, t)$ is called the "matrizant" (or "fundamental solution matrix" or "state transition matrix") of the system (2). Since each of its columns satisfies (2), it must itself satisfy

$$(4) \qquad \Omega' = A\Omega,$$

where

$$\Omega(t_0, t_0) = I.$$

If the function $\delta \mathbf{X}(t)$ were to have initial conditions $\delta \mathbf{X}(t_0) = \delta \mathbf{X}_0$, then the appropriate solution of (2) would be

(5) $\delta \mathbf{X}(t) = \Omega(t_0, t)\,\delta \mathbf{X}_0.$

It is clear that $\Omega(t_0, t)$ is the Jacobian matrix with components $\partial X_i / \partial X_{0,j}$, etc.

Consider the relations between residuals $\delta \mathbf{X}$ at times t_0, t_1, and t_2. We have

$$\delta \mathbf{X}_2 = \Omega(t_0, t_2)\,\delta \mathbf{X}_0,$$

and

$$\delta \mathbf{X}_2 = \Omega(t_1, t_2)\,\delta \mathbf{X}_1$$
$$= \Omega(t_1, t_2)\,\Omega(t_0, t_1)\,\delta \mathbf{X}_0.$$

Therefore

(6) $\Omega(t_0, t_2) = \Omega(t_1, t_2)\,\Omega(t_0, t_1),$

a result that is also evident from the fact that $\Omega(t_0, t)$ is a Jacobian matrix.

Consider the application of the matrizant to some situations in the context of astronautics. Suppose that a reference orbit has been calculated. If some maximum permissible error at time t_1 is specified, then the maximum permissible error at any earlier time t_0 can be calculated if $\Omega(t_1, t_0)$ is known. If an error is observed at t_0, the effect at a later time t_1 can be calculated using $\Omega(t_0, t_1)$. But if t_1 is fixed and t_0 varies it is obviously inconvenient to solve the equations for $\Omega(t_0, t_1)$ many times for different t_0, and it is better to put

(7) $\delta \mathbf{X}_1 = \Omega^{-1}(t_1, t_0)\,\delta \mathbf{X}_0,$

and solve the corresponding equations with the initial conditions applied at t_1. We notice, incidentally, that $\Omega^{-1}(t_1, t_0) = \Omega(t_0, t_1)$. Furthermore, it is possible to avoid the inversion of the matrix; for let

$$\Upsilon(t_1, t)\,\Omega(t_1, t) = I.$$

Differentiating with respect to t, and using (4), we find

(8) $\Upsilon' = -\Upsilon A.$

Equation (8) is called the "adjoint equation" of (4). (The use of the adjoint equations in this sort of context was first cultivated in ballistics, and is described in [1].)

Now suppose that equations (4) and (8) have both been solved, the initial conditions making each matrix equal to the identity matrix at time t_1. Writing (6) as

(6A) $\qquad \Omega(t_0, t_2) = \Omega(t_1, t_2)\, \Omega^{-1}(t_1, t_0) = \Omega(t_1, t_2)\, \Upsilon(t_1, t_0),$

we see that the matrizant relating *any* two times can be found.

Normally $\mathbf{X}$ will have six components, of position

$$\mathbf{r} = \begin{bmatrix} x \\ y \\ z \end{bmatrix} = \begin{bmatrix} X_1 \\ X_2 \\ X_3 \end{bmatrix}, \text{ and velocity } \mathbf{r}' = \begin{bmatrix} x' \\ y' \\ z' \end{bmatrix} = \begin{bmatrix} X_4 \\ X_5 \\ X_6 \end{bmatrix}.$$

Let the matrizant in (5) be subdivided into four three-by-three matrices:

(9) $\qquad\qquad \Omega(t_0, t) = \begin{bmatrix} U(t_0, t) & V(t_0, t) \\ W(t_0, t) & Y(t_0, t) \end{bmatrix}.$

Suppose that an error $\delta \mathbf{r}_0$ is observed at t_0, and it is required that *after* a thrust has been applied there will be a velocity residual $\delta \mathbf{r}_0'$ such that $\delta \mathbf{r}$ at time t is zero. Then we have

(10) $\qquad\qquad\qquad\qquad \delta \mathbf{r}_0' = -\, V^{-1} U \delta \mathbf{r}_0.$

Consider motion subject to a force function R. The differential equations of motion are

$$X_1' = X_4,\ X_2' = X_5,\ X_3' = X_6,\ X_4' = \partial R/\partial X_1,$$

$$X_5' = \partial R/\partial X_2,\ X_6' = \partial R/\partial X_3.$$

The first variational equations can be written as $\delta \mathbf{X}' = A \delta \mathbf{X}$, where

(11)
$$A = \begin{bmatrix} 0 & 0 & 0 & 1\ 0\ 0 \\ 0 & 0 & 0 & 0\ 1\ 0 \\ 0 & 0 & 0 & 0\ 0\ 1 \\ \partial^2 R/\partial X_1^2 & \partial^2 R/\partial X_1 \partial X_2 & \partial^2 R/\partial X_1 \partial X_3 & 0\ 0\ 0 \\ \partial^2 R/\partial X_1 \partial X_2 & \partial^2 R/\partial X_2^2 & \partial^2 R/\partial X_2 \partial X_3 & 0\ 0\ 0 \\ \partial^2 R/\partial X_3 \partial X_1 & \partial^2 R/\partial X_3 \partial X_2 & \partial^2 R/\partial X_3^2 & 0\ 0\ 0 \end{bmatrix}$$

$$= \begin{bmatrix} 0 & I \\ B & 0 \end{bmatrix}, \text{ say.}$$

Then substituting from (9) and (11) into the equation $\Omega' = A\Omega$, we find

$$U' = W, \quad V' = Y, \quad W' = BU, \quad Y' = BV,$$

from which

(12) $$\qquad\qquad U'' = BU \text{ and } V'' = BV.$$

Equations (12) are to be solved subject to the initial conditions

(13)
$$\begin{aligned} U(t_0, t_0) &= I, \quad V(t_0, t_0) = 0; \\ U'(t_0, t_0) &= 0, \quad V'(t_0, t_0) = I. \end{aligned}$$

The columns of U and V are six linearly independent solutions of the equation

(14) $$\qquad\qquad \delta\mathbf{r}'' = B\delta\mathbf{r},$$

which is the first variational equation of the equation of motion in the form $\mathbf{r}'' = \operatorname{grad} R$.

Let $\delta\mathbf{r}_0$ and $\delta\mathbf{r}_0'$ be the initial increments in position and velocity to be applied to the reference orbit at time t_0. Then at time t

(15) $$\qquad\qquad \delta\mathbf{r} = U(t_0, t)\,\delta\mathbf{r}_0 + V(t_0, t)\,\delta\mathbf{r}_0'.$$

$\delta\mathbf{r}$ is a solution of (14), and since the components of $\delta\mathbf{r}_0$ and $\delta\mathbf{r}_0'$ can be considered as six independent, arbitrary constants, it is clear that (15) is the general solution of (14).

The matrizant and its components should always be considered as functions of two variables (two independent times). Now consider

$$Z(t_0, t) = \int_{t_0}^{t} U(\tau, t)\, d\tau.$$

We have

$$\partial Z/\partial t = I + \int_{t_0}^{t} (\partial U(\tau, t)/\partial t)\, d\tau$$

and

$$\partial^2 Z/\partial t^2 = \int_{t_0}^{t} (\partial^2 U(\tau, t)/\partial t^2)\, d\tau$$

$$= \int_{t_0}^{t} B(t)\, U(\tau, t)\, d\tau = BZ.$$

So Z satisfies the differential equation as well as the initial conditions for V, and must therefore be identical with V. Hence

(16) $$U(t_0, t) = - \partial V(t_0, t)/\partial t_0.$$

So the matrizant (9) can be written

(17) $$\Omega(t_0, t) = \begin{bmatrix} - \partial V/\partial t_0 & V(t_0, t) \\ - \partial^2 V/\partial t \partial t_0 & \partial V/\partial t \end{bmatrix}.$$

Let

$$\Omega^{-1}(t_0, t) = \Upsilon(t_0, t) = \begin{bmatrix} \overline{U} & \overline{V} \\ \overline{W} & \overline{Y} \end{bmatrix}.$$

Then from (8) and (11) we find

$$\begin{bmatrix} \overline{U}' & \overline{V}' \\ \overline{W}' & \overline{Y}' \end{bmatrix} = - \begin{bmatrix} \overline{U} & \overline{V} \\ \overline{W} & \overline{Y} \end{bmatrix} \begin{bmatrix} 0 & I \\ B & 0 \end{bmatrix}.$$

Therefore

$$\overline{V}' = - \overline{U}, \quad \overline{Y}' = - \overline{W}, \quad \overline{U}' = - \overline{V}B, \quad \overline{W}' = - \overline{Y}B.$$

So

(18) $$\overline{V}'' = \overline{V}B, \qquad \overline{Y}'' = \overline{Y}B,$$

where

(19) $$\begin{aligned} \overline{V}(t_0, t_0) = 0, \qquad \overline{Y}(t_0, t_0) = I, \\ \overline{V}'(t_0, t_0) = - I, \quad \overline{Y}'(t_0, t_0) = 0. \end{aligned}$$

Now B is symmetrical, so that transposing equations (18), we find

(18A) $$\overline{V}''^{T} = B \overline{V}^{T}, \quad \overline{Y}''^{T} = B \overline{Y}^{T}.$$

Comparing (18A) and (19) with (12) and (13), we see that

$$\overline{V} = - V^{T}, \quad \overline{U} = Y^{T}, \quad \overline{W} = - W^{T}, \text{ and } \overline{Y} = U^{T}.$$

Therefore

(20) $$\Omega^{-1}(t_0, t) = \begin{bmatrix} U & V \\ W & Y \end{bmatrix}^{-1} = \begin{bmatrix} Y^{T} & - V^{T} \\ - W^{T} & U^{T} \end{bmatrix},$$

a result that applies in general when the original system of equations is canonical.

The components of a matrizant would normally have to be

found numerically; but in some cases it is possible to find them analytically. This is notably so in the case of Keplerian motion, for which the components of V are given in a paper in [2]. This matrizant has possible applications in perturbation problems in celestial mechanics. The components are most easily found, not by solving the differential equation, but by considering, from first principles, what the effects of errors in velocity at time t_0 will be on errors in position at time t.

Consider the equation

(21) $$\delta X' = A\,\delta X + g(t),$$

in which a "forcing function", $g(t)$, has been added to (2). The equation is no longer homogeneous, and one way to solve it is to take the solution of the homogeneous part, viz.,

(22) $$\delta X(t) = \Omega(t_0, t)\,\delta X_0,$$

and allow the arbitrary constants, δX_0, to vary. Then

$$\delta X' = \Omega'\delta X_0 + \Omega\delta X_0'$$
$$= A\Omega\delta X_0 + \Omega\delta X_0'.$$

Substituting this, and (5), into (21), we have

$$A\Omega\delta X_0 + \Omega\delta X_0' = A\Omega\delta X_0 + g,$$

so that

$$\delta X_0 = \int_{t_0}^{t} \Omega^{-1}g\,dt.$$

The complete and general solution is therefore

(23) $$\delta X = \Omega(t_0, t)\,\delta X_0 + \Omega(t_0, t)\int_{t_0}^{t}\Omega^{-1}(t_0, \tau)\,g(\tau)\,d\tau,$$

where δX_0 is once again constant. This is the exact solution of (21), subject to the initial conditions $\delta X(t_0) = \delta X_0$; no conditions about orders of magnitude are imposed. The first term, which includes the arbitrary constants, is the complementary function, and the second is the particular integral. If the particular integral is to be found numerically, probably the best procedure is to solve equations (21) subject to the initial conditions $\delta X(t_0) = 0$. (23) can be simplified by the use of the multiplication formula (6) to give

$$(24) \qquad \delta \mathbf{X} = \Omega(t_0, t) \, \delta \mathbf{X}_0 + \int_{t_0}^{t} \Omega(\tau, t) \, \mathbf{g}(\tau) \, d\tau.$$

Another form is

$$(25) \qquad \delta \mathbf{X} = \Omega(t_0, t) \, \delta \mathbf{X}_0 + \Omega(s, t) \int_{t_0}^{t} \Omega(\tau, s) \, \mathbf{g}(\tau) \, d\tau,$$

where s is an arbitrary time, chosen to make Ω as simple as possible.

If the equations of motion are in cartesian coordinates, then the first three components of $\mathbf{g}$ are zero; writing the last three as $\mathbf{f}$, we have, from (24)

$$(26) \qquad \delta \mathbf{r} = U(t_0, t) \, \delta \mathbf{r}_0 + V(t_0, t) \, \delta \mathbf{r}_0' + \int_{t_0}^{t} V(\tau, t) \, \mathbf{f}(\tau) \, d\tau.$$

Or, from (25) and (20),

$$(27) \qquad \begin{aligned} \delta \mathbf{r} = {} & U(t_0, t) \, \delta \mathbf{r}_0 + V(t_0, t) \, \delta \mathbf{r}_0' \\ & + \left[\, U(s, t) \quad V(s, t) \,\right] \int_{t_0}^{t} \begin{bmatrix} -\, V^T(s, \tau) \\ U^T(s, \tau) \end{bmatrix} \mathbf{f}(\tau) \, d\tau. \end{aligned}$$

In the case of disturbed Keplerian motion, s would certainly be a time of perihelion passage. Also in this case there are advantages in changing the independent variable from the time to the eccentric anomaly in the reference orbit.

References

1. G. A. Bliss, *Mathematics for exterior ballistics*, Wiley & Sons, New York, 1944.
2. J. M. A. Danby, *Integration of the equations of planetary motion in rectangular coordinates*, Astronom. J. **67** (1962), 287-299.

YALE UNIVERSITY OBSERVATORY

Boris Garfinkel

The Lagrange-Hamilton-Jacobi Mechanics

I. **Introduction.** Consider a dynamical system specified by a set of *generalized* coordinates $q = (q_i)$ $(i = 1, 2, \cdots, N)$, subject to m holonomic constraints, of the form

$$(1) \qquad\qquad \phi(q, t) = 0,$$

where $\phi = (\phi_j)$ $(j = 1, 2, \cdots, m)$. Then m of the coordinates can be eliminated; the remaining coordinates number

$$(2) \qquad\qquad n = N - m,$$

which equals the number of degrees of freedom. Such a system is said to be *holonomic*. If the constraint equations (1) do not contain the time explicitly, we say the system is *scleronomic*, in contrast to a *rheonomic* system, with time-dependent contraints.

LEMMA 1. *The kinetic energy is a quadratic function of the generalized velocities $\dot{q}$.*

PROOF. Let the system be composed of N particles of mass m_k and position vectors r_k of the form

$$(3) \qquad\qquad r_k = r_k(q, t); \qquad k = 1, 2, \cdots, N.$$

40

Then

(4) $$\dot{r}_k = r_{k,t} + r_{k,q_i}\dot{q}_i,$$

(5) $$\dot{r}_k^2 = r_k^2 + 2r_{k,t}\cdot r_{k,q_i}\dot{q}_i + r_{k,q_i}\cdot r_{k,q_j}\dot{q}_i\dot{q}_j, \qquad i,j = 1,2,\cdots,n,$$

where the subscripts t and q denote the arguments of partial differentiation, and summation with respect to repeated indices i,j is understood. The substitution from (4) and (5) into the defining equation

(6) $$T = \frac{1}{2}m_k\dot{r}_k^2$$

leads to an expression of the form

(7) $$T = \frac{1}{2}\vec{q}G\dot{q} + \bar{b}\dot{q} + \frac{1}{2}c.$$

Here G is the square matrix,

(8) $$G = (m_k r_{k,q_i}\cdot r_{k,q_j}),$$

b is a column matrix,

$$b = (m_k r_{k,t}\cdot r_{k,q_i}),$$

and c is a scalar,

(9) $$c = m_k r_{k,t}^2.$$

The bar over a matrix will denote its transpose; thus $\bar{b}$ is a row matrix. The conclusion now follows by inspection of (7).

COROLLARY. *If the system is scleronomic, the kinetic energy is a pure quadratic function of the generalized velocities.*

PROOF. Since t is absent in (3), it follows that $r_{k,t} = 0$, so that $b = 0$, $c = 0$, and

(10) $$T = \frac{1}{2}\vec{q}\,G\dot{q}.$$

Observe that: (1) G is the *metric tensor* of the *configuration space*; i.e., the space of the coordinates q; (2) G is symmetric and positive-definite; i.e.,

(11) $$\bar{G} = G, \qquad |G| > 0.$$

The differential of work dW can be written

(12) $$dW = \bar{Q}\,dq,$$

where $Q = (Q_i)$ is the *generalized force*. We say the system is *monogenic* if there exists a potential function of the form

(13) $$V = V(q, \dot{q}; t),$$

such that

(14) $$Q = \frac{d}{dt} V\dot{q} - V_q.$$

We note that if the potential is not velocity-dependent, then $V\dot{q} = 0$, and

(15) $$Q = -V_q;$$

i.e., the force equals the negative *gradient* of V in the configuration space. A system is said to be *conservative* if

(16) $$V = V(q).$$

Generally, we define the *Lagrangian* function L by

(17) $$L(q, \dot{q}, t) \equiv T - V.$$

An axiomatic foundation of analytical dynamics is furnished by the Hamilton Principle. The latter asserts that

(18) $$\delta \int_{t_1}^{t_2} L \, dt = 0$$

on a dynamical path connecting two fixed points $q(t_1)$ and $q(t_2)$ in the (q, t) space. If the system is holonomic, it is then necessary that L satisfy the Euler equations of the calculus of variations;

(19) $$\frac{d}{dt} L_{\dot{q}} = L_q.$$

Define the set of *conjugate momenta* p by

(20) $$p = L_{\dot{q}}(q, \dot{q}, t).$$

The latter equation can be solved for $\dot{q}$ to yield

(21) $$\dot{q} = \dot{q}(q, p, t)$$

provided

(22) $$|L_{\dot{q}\dot{q}}| \neq 0.$$

LEMMA 2. *If a system is scleronomic and conservative, then* $|L_{qq}| \neq 0$.

PROOF. From (10), (11), and (16) we deduce

$$L = \frac{1}{2}\bar{\dot{q}}G\dot{q} - V(q),$$

(23)
$$L_q = G\dot{q},$$

$$L_{qq} = G,$$

$$|L_{qq}| = |G| > 0.$$

The Hamiltonian H of a system is defined by

(24) $$H(q,p,t) \equiv \bar{p}\dot{q} - L(q,\dot{q},t),$$

where $\dot{q}$ appearing in the right-hand member is to be expressed as a function of q,p,t by means of (21).

THEOREM 1. *If the system is scleronomic, and if the potential is velocity independent, then the Hamiltonian equals the total energy E of the system.*

PROOF. From (23.1) and (20) it follows that

(25)
$$p = G\dot{q},$$

$$\dot{q} = G^{-1}p.$$

Here we recall that G is symmetric and that the transpose of the product of several matrices equals the product of the transposed matrices arranged in the reverse order. Therefore, with the aid of (10), we obtain $T = \frac{1}{2}\bar{\dot{q}}G\dot{q} = \frac{1}{2}\bar{p}G^{-1}GG^{-1}p$, leading to

(26) $$T = \frac{1}{2}\bar{p}G^{-1}p.$$

Finally, (24) and (17) imply $H = 2T - L$ and

(27) $$H = T + V \equiv E.$$

A set of $2n$ variables (q,p) is said to be *canonical* if it satisfies the *canonical equations* of Hamilton,

(28) $$\dot{q} = H_p, \qquad \dot{p} = -H_q.$$

We shall show that (28) is implied by (18). In the original formulation of the Hamilton Principle, the end points were fixed in the configuration space. We shall now modify the Principle by fixing the end points in the *phase-space* of (q,p).

THEOREM 2. *The Hamilton Principle implies the canonical equations of motion.*

PROOF. In view of (24), the integrand of (18) can be written as

$$(29) \qquad\qquad L = \bar{p}\dot{q} - H(q,p,t), \qquad\qquad .$$

and can be regarded as a function of (q,p,t) and the derivatives $(\dot{q},\dot{p})$. Then the hypothesis requires that $L(q,p,\dot{q},\dot{p},t)$ satisfy the Euler equations in the form

$$(30) \qquad\qquad \frac{d}{dt} L_{\dot{q}} = L_q, \qquad \frac{d}{dt} L_{\dot{p}} = L_p.$$

From (20) and (29), we deduce

$$(31) \qquad L_{\dot{q}} = p, \quad L_q = -H_q, \quad L_{\dot{p}} = 0, \quad L_p = \dot{q} - H_p,$$

which together with (30) implies (28).

THEOREM 3. *If the Hamiltonian does not contain the time explicitly, then it is a constant of the motion.*

PROOF. With the aid of (28),

$$(32) \qquad\begin{aligned} \dot{H} &= H_t + \dot{\bar{q}}H_q + \dot{\bar{p}}H_p \\ &= H_t + \bar{H}_p H_q - \bar{H}_q H_p \\ &= H_t, \end{aligned}$$

since the *inner product* of the vectors H_q and H_p is commutative. The hypothesis $H_t = 0$ then implies $\dot{H} = 0$ and $H = \text{const}$.

Let the state of a system in the phase-space be specified by a $2n$-vector

$$(33) \qquad\qquad x = \begin{pmatrix} q \\ p \end{pmatrix},$$

and let D be the corresponding $2n$-vector differential operator. We introduce the *canonical matrix* Φ_0 represented by

$$(34) \qquad\qquad \Phi_0 = \begin{pmatrix} O & I \\ -I & O \end{pmatrix},$$

where O denotes the $n \times n$ *null-matrix,* and I the $n \times n$ *identity matrix,* and note the following identities:

(35) $$\overline{\Phi}_0 = -\Phi_0 = \Phi_0^{-1}, \qquad \Phi_0^2 = -I, \qquad |\Phi_0| = 1;$$

i.e., Φ_0 is *skew-symmetric* and *orthogonal.* With the definitions (33) and (34), the canonical equations (28) assume the form

(36) $$\dot{x} = \Phi_0 H_x = \Phi_0 DH.$$

The *extended* phase-space of $2n + 2$ dimensions includes the conjugate variables

(37) $$q_0 \equiv t, \quad p_0 \equiv -H.$$

With the *extended* Hamiltonian, defined by

(38) $$\mathscr{H} \equiv H + p_0,$$

the canonical equations,

(39) $$\dot{q}_0 = \mathscr{H}_{p_0}, \quad \dot{p}_0 = -\mathscr{H}_{q_0},$$

are, indeed, satisfied. The first one is equivalent to the trivial identity $1 = 1$; the second is equivalent to $\dot{H} = H_t$, which is equation (32). The *invariant imbedding* described in this paragraph eliminates the time from the Hamiltonian, and reduces t to the role of a parameter.

II. **Transformation of variables.** Consider the transformation $x \to x'$ defined by

(40) $$x' = x'(x).$$

If the transformation should involve the time, the latter could be eliminated by the imbedding construction of §I. Let J denote the Jacobian matrix of the transformation,

(41) $$J \equiv x'_x = (\partial x'_i / \partial x_j), \qquad i,j = 1, 2, \cdots, 2n.$$

The following properties of the Jacobian will be used:
   (1) $d = d\overline{x}D$, where d is the total differential operator;
   (2) $\overline{J} = D\overline{x}'$;
   (3) $dx' = J dx$;
   (4) $a_c = a_b b_c$;
   (5) $a_a = I$;
   (6) $a_b = (b_a)^{-1}$;
   (7) $D' = \overline{J}^{-1} D$.

The proofs are quite elementary and will be omitted here. In the language of tensor analysis, we say that dx is a *contravariant* vector, whose transformation is described by the matrix J in (3), while D is a *covariant* vector, transforming by the matrix J^{-1} in (7). The two matrices so related that the transpose of one equals the inverse of the other are said to be mutually *contragredient*.

THEOREM 4. *If J is the Jacobian matrix of the transformation $x'(x)$ on a canonical set x with the Hamiltonian H, then the equations of motion in terms of x' are*

$$(42) \qquad\qquad \dot{x}' = \Phi D' H,$$

where

$$(43) \qquad\qquad \Phi \equiv J \Phi_0 \bar{J},$$

and the scalar invariant H is expressed as a function of x'.

The proof involves the relations $\dot{x}' = J\dot{x}$ and $D = \bar{J}D'$ implied by the properties (3) and (7) respectively, and the use of the canonical equation (36).

The $2n \times 2n$ matrix Φ defined by (43) is named after Poisson; its elements Φ_{ij} are the Poisson *brackets*, denoted by the symbol (i, j) in the present chapter.

III. **Canonical transformation.** A transformation is said to be *canonical* if it preserves the canonical form of the equations of motion.

THEOREM 5. *A transformation with the Jacobian J is canonical if and only if*

$$(44) \qquad\qquad J\Phi_0 \bar{J} = \Phi_0.$$

The proof follows from (36), (42), and (43). The Jacobi relation (44) is equivalent to the following:

$$(45) \qquad \begin{aligned} q_q' \, \bar{q}_p' - q_p' \, \bar{q}_q' &= O, \\ q_q' \, \bar{p}_p' - q_p' \, \bar{p}_q' &= I, \\ p_q' \, \bar{p}_p' - p_p' \, \bar{p}_q' &= O. \end{aligned}$$

The proof involves the *block representations* (34) and

$$(46) \qquad\qquad J = \begin{pmatrix} q_q' & q_p' \\ p_q' & p_p' \end{pmatrix},$$

with the use of the rule of matrix *block-multiplication*.

Another form of (44) is

$$(47) \qquad\qquad \Phi_0 \overline{J} = J^{-1} \Phi_0,$$

which is equivalent to

$$(48) \qquad\qquad \begin{aligned} \overline{q}_q' &= p_{p'}, & \overline{q}_p' &= -q_{p'}, \\ \overline{p}_q' &= -p_{q'}, & \overline{p}_p' &= q_{q'}. \end{aligned}$$

THEOREM 6. *For any function $S(q, p', t)$ with $|S_{qp'}| \neq 0$, the transformation $(q, p) \to (q', p')$ defined by*

$$(49) \qquad\qquad p = S_q, \qquad q' = S_{p'}, \qquad H' = H + S_t$$

is a canonical transformation.

PROOF. Let (q, p) be canonical variables satisfying (28) with the Hamiltonian H. Define L and L' by

$$(50) \qquad\qquad \begin{aligned} L\,dt &= \overline{p}\,dq - H\,dt, \\ L'\,dt &= \overline{p}'\,dq' - H'\,dt. \end{aligned}$$

Then it follows with the aid of (49) that

$$\begin{aligned} L\,dt - L'\,dt &= \overline{S}_q\,dq + \overline{S}_{p'}\,dp' + S_t\,dt - d(\overline{p}'\,q') \\ (51) \qquad\qquad &= dS - d(\overline{p}'\,q') \\ &= dW, \end{aligned}$$

where

$$(52) \qquad\qquad W \equiv S - \overline{p}'\,q'.$$

Hence

$$(53) \qquad\qquad \delta \int_{t_1}^{t_2} (L - L')\,dt = \delta W \Big|_{t_1}^{t_2} = 0,$$

since W is a function of q, p', t and since the end points are assumed fixed in phase-space. On the other hand,

$$(54) \qquad\qquad \delta \int_{t_1}^{t_2} L\,dt = 0$$

in consequence of (28). Hence $\delta \int_{t_1}^{t_2} L'\,dt = 0$, or

(55) $$\delta \int_{t_1}^{t_2} (\overline{p}' \, \dot{q}' - H')dt = 0,$$

which implies

(56) $$\dot{q}' = H'_p, \qquad p' = - H_{q'}$$

by the argument of Theorem 2. Provided $|S_{qp'}| \neq 0$, the first equation of (49) can be solved for p' and the result substituted into the second equation. Then the transformation is specified by the equations of the form $q' = q'(q,p,t)$ and $p' = p'(q,p,t)$.

COROLLARY 1. *The transformation is canonical if*

(57) $$\overline{p} \, dq - H \, dt - (\overline{p}' \, dq' - H' \, dt) = dW,$$

where dW is a total differential.

COROLLARY 2. *The transformation is canonical if $H' = H$ and*

(58) $$\overline{p} \, dq - \overline{p}' \, dq' = 0.$$

Of the $2n$ arguments of S, n are old variables and n are new. Altogether four forms of S are distinguished, which are related to W as indicated below:

(59)
$$
\begin{array}{llll}
(1) \ W = S(q,q'), & \qquad & p = S_q, & \qquad p' = - S_{q'}; \\
(2) \ W = S(q,p') - \overline{p}' \, q', & & p = S_q, & \qquad q' = S_{p'}; \\
(3) \ W = - S(q',p) + \overline{p}q, & & q = S_p, & \qquad p' = S_{q'}; \\
(4) \ W = S(p,p') - \overline{p}' \, q' + \overline{p} \, q, & & q = - S_p, & \qquad q' = S_{p'}.
\end{array}
$$

IV. **Some algebraic properties of a canonical transformation.** A transformation such that q' is a function of q only and p' a function of p only is termed an *extended point transformation.*

THEOREM 7. *An extended point transformation is canonical if and only if (1) it is linear, and (2) the coordinate and the momentum transformations are described by mutually contragredient matrices.*

PROOF. Since $q' = q'(q)$ and $p' = p'(p)$, it follows from (48.2) and property (6) of §II that

(60) $$q'_q = (\overline{p}'_p)^{-1} \equiv A,$$

where the matrix A is independent of q and p.

COROLLARY 1. *A linear transformation of the form*

(61) $$q' = Aq, \qquad p' = Bp$$

is canonical if and only if the matrices A and B satisfy the relation

(62) $$A = \overline{B}^{-1}.$$

In the proof we note that

(63) $$q'_q = A, \qquad p'_p = B.$$

Then (62) follows from (60).

COROLLARY 2. *If the coordinates are multiplied by a constant factor* λ, *while the momenta are divided by the same factor, the transformation is canonical.*

The proof follows from (62), which is clearly satisfied by

(64) $$A = \lambda I, \qquad B = I/\lambda.$$

THEOREM 8. *The set of all canonical transformations is a group.*

In the proof, the properties of closure, associativity, the existence of the identity element, and the existence of the inverse are established with the aid of (44), the rules of matrix algebra, and the properties of the Jacobian, numbered 4-6 in §II. A useful identity,

(65) $$J \Phi_0 J = \Phi_0,$$

is derived from (44) by premultiplying the latter by $\Phi_0 J^{-1}$ and postmultiplying by $\Phi_0 J$.

THEOREM 9. *The Jacobian determinant of a canonical transformation equals unity.*

PROOF. Since $|\Phi_0| = 1$, it follows from (44) that $|J| = \pm 1$. The negative sign is then ruled out by two circumstances: (1) the group of transformations is continuous; (2) for the identity transformation $x' = x$ the relations $J = I$ and $|J| = +1$ hold.

V. The Hamilton-Jacobi equation.

THEOREM 10. *If the generating function* $S(q, p', t)$ *of a canonical transformation is so chosen that the new Hamiltonian H' is identically zero, then the new coordinates q' and the new momenta p' are constants of the motion, and S satisfies the Hamilton-Jacobi partial differential equation*

(66) $H(q, S_q, t) + S_t = 0.$

PROOF. The first part of the conclusion follows from (28) with all quantities primed and $H' \equiv 0$; (66) is a direct consequence of (49.3).

THEOREM 11. *If the Hamilton-Jacobi equation (66) has a solution $S(q, \alpha, t)$ with α a set of n constants, then the functions $q(t)$ determined by the equations $S_\alpha = \beta$, with β another set of n constants, and the function $p(t)$ determined by the equations $p = S_q$, are the coordinates and the momenta respectively in the solution of the dynamical problem with the Hamiltonian $H(q, p, t)$.*

PROOF. Let the new variables be defined by

(67) $p' = \alpha, \qquad q' = \beta.$

By Theorem 6, $S(q, \alpha, t)$ is then the generating function of the canonical transformation $(q, p) \to (\beta, \alpha)$ with the new Hamiltonian $H' = H + S_t = 0$, vanishing identically in virtue of (66). Since α and β trivially satisfy the canonical equations with $H' \equiv 0$, they are canonical variables. Furthermore, $(\beta, \alpha) \to (q, p)$, being the inverse of a canonical transformation, is also a canonical transformation by Theorem 8. Therefore, (q, p) are canonical variables with the Hamiltonian H. If α and β are so chosen as to satisfy the initial conditions $q(0) = q_o$ and $p(0) = p_o$, the solution of the dynamical problem is unique.

Thus, if one succeeds in solving the Hamilton-Jacobi equation to obtain a function $S(q, \alpha, t)$, he has in effect performed a canonical transformation $(q, p) \to (\beta, \alpha)$ with $H' \equiv 0$. In the new variables, the phase path shrinks to a point $\alpha = $ const. $\beta = $ const. The old variables are then expressed with the aid of the function S in the form $q(\alpha, \beta, t)$ and $p(\alpha, \beta, t)$. The detailed procedure is summarized below:

(1) Construct the Hamiltonian $H(q, p, t)$ of the system and set up the Hamilton-Jacobi equation (66).

(2) Find a solution $S(q, \alpha, t)$ of (66) containing n arbitrary constants α.

(3) Write the kinematic equations of motion $S_\alpha = \beta$, where β is another set of n arbitrary constants.

(4) Solve the latter equations for q, obtaining $q = q(\alpha, \beta, t)$.

(5) Determine the constants of integration α, β from the initial

conditions $q(0)$, $\dot{q}(0)$.

The Hamilton-Jacobi equation can be solved exactly if it is *separable*.

VI. **The separable case.** A system is said to be partially separable if a solution of Hamilton-Jacobi equation can be represented by

(68) $$S(q, t) = W(q) + \mathscr{T}(t),$$

where W is a function of q only, called the Hamilton Principal Function, and $\mathscr{T}$ a function of t only.

THEOREM 12. *A conservative scleronomic system is partially separable.*

PROOF. Since t is absent in the Hamiltonian, (66) assumes the form

(69) $$H(q, S_q) + S_t = 0,$$

which separates into

(70) $$\begin{aligned} H(q, W_q) &= \alpha_1 = \text{const.}, \\ \mathscr{T}' &= -\alpha_1 \end{aligned}$$

by means of the substitution (68). Then (70.2) leads to $\mathscr{T} = -\alpha_1 t$ and

(71) $$S = W(q, \alpha) - \alpha_1 t,$$

where W is to be obtained as a solution of (70.1) with n arbitrary constants α.

A coordinate is said to be *ignorable* if it does not appear explicitly in the Hamiltonian. By the argument of the proof of Theorem 12, such a coordinate q_i contributes to S a linear term $\alpha_i q_i$. In Theorem 12, the role of ignorable coordinate is assumed by the time. Another example is furnished by the azimuth coordinate ϕ of a particle in an axi-symmetric potential field.

We note that α_1 in (69) is the total energy by Theorem 1. The transformation equations are $p = W_q$, and

(72) $$\begin{aligned} W_{\alpha_1} &= \beta_1 + t, \\ W_{\alpha_2} &= \beta_2, \\ &\;\cdots\cdots\cdots \end{aligned}$$

A system is said to be *completely separable* if a solution of the

Hamilton-Jacobi equation can be represented in the form

$$S(q,t) = \sum_0^n S_i(q_i),$$

(73)

$$q_0 \equiv t.$$

Then $p = S_q$ leads to

(74) $\qquad\qquad S_i' = p_i(q_i, \alpha), \qquad (i = 0, 1, \cdots, n),$

so that each momentum is a function of the conjugate coordinate and the n *separation constants* α. In particular, for $i = 0$, $S_0' = p_0 = -\alpha_1 = -H$ by the argument of Theorem 12. This result is in accord with (37). In the process of solution, the Hamilton Jacobi equation splits into $n + 1$ ordinary differential equations (74). With the aid of (73), S is obtained by quadrature,

(75) $\qquad\qquad S = \sum_{i=0}^n \int p_i(q_i, \alpha) dq_i = \int (\overline{p}\, dq - H\, dt).$

In view of (29) and (18), the latter expression can be identified with the *action*, defined by

(76) $\qquad\qquad\qquad S = \min \int L\, dt.$

A useful criterion for separability is furnished by the *Staekel Condition*: In an orthogonal coordinate system q, there exists a nonsingular matrix $\Phi_{ij}(q_i)$ and a vector $\psi_i(q_i)$ with $i, j = 1, 2, \cdots, n$ such that

$$V = \sum_1^n \psi_i / g_{ii},$$

(77)

$$(\Phi^{-1})_{1j} = 1/g_{jj},$$

where g_{ii} are the elements of the diagonal matrix G, representing the metric tensor of the configuration space.

THEOREM 13. *If a conservative scleronomic system satisfies the Staekel Condition, then the system is separable.*

PROOF. In view of Theorem 1 and (26), the Hamiltonian can be written

$$(78) \qquad H = \frac{1}{2}\, \overline{p}G^{-1}(q)p + V(q).$$

The Hamilton-Jacobi equation (70.1) then becomes

$$(79) \qquad \frac{1}{2}\, \overline{W}_q G^{-1}(q)\, W_q + V(q) = \alpha_1.$$

The substitution

$$(80) \qquad W = \sum_1^n W_i(q_i)$$

converts (79) into

$$(81) \qquad \frac{1}{2}\sum_1^n (W_i')^2/g_{ii} + V = \alpha_1.$$

We shall show that if the functions W_i are defined by

$$(82) \qquad (W_i')^2 = -2\psi_i + \sum_j \Phi_{ij}\alpha_j, \qquad i,j = 1,2,\cdots,n,$$

where α_i are the separation constants, then (81) is identically satisfied. Let a and b denote the n-vectors defined by

$$(83) \qquad a = ((W_i')^2), \qquad b = (g_{ii}^{-1}).$$

In the matrix notation (81) and (82) then become

$$(84) \qquad \begin{aligned} &\frac{1}{2}\,\overline{b}a + V = \alpha_1, \\ &a = -2\psi + 2\Phi\alpha. \end{aligned}$$

By the hypothesis (77),

$$(85) \qquad \overline{b}\psi = V, \qquad (1,0,\cdots)\Phi^{-1} = \overline{b}.$$

Finally, the elimination of a and b reduces (84.1) to $(1,0,\cdots)\Phi^{-1}\Phi\alpha = \alpha_1$, which is readily seen to be an identity.

Examples of a Staekel system for a particle are furnished by the following potentials:

(1) $V = -1/r$, where r is a spherical coordinate.

(2) $V = -1/r + 3\, J_2[\, c_1(\cos^2\theta - c_2)/2r^2 + c_3/r + c_4/r^3\,]$,

where r, θ, ϕ are spherical coordinates and c_1, c_2, c_3, c_4 are disposable parameters. In the artificial satellite theory, $c_3 = 0$ was used by Sterne in [8], and $c_4 = 0$ by Garfinkel in [9].

(3) $V = -\xi/(\xi^2 + \eta^2)$, where ξ, η, ϕ are spheroidal coordinates, was used in the artificial satellite theory by Vinti in [12].

(4) The most general separable potential in spherical coordinates is of the form

(86) $$V = f_1(r) + f_2(\theta)/r^2 + f_3(\phi)/r^2 \sin^2\theta.$$

The Staekel Condition is satisfied with

$$\psi_i = f_i,$$

(87) $$\Phi = \begin{pmatrix} 1 & -r^{-2} & 0 \\ 0 & 1 & \csc^2\theta \\ 0 & 0 & 1 \end{pmatrix}.$$

VII. **Conditional periodicity.** For a Staekel system, (74) and (71) imply

(88) $$p_i = W_i'(q_i, \alpha); \quad (i = 1, 2, \cdots, n).$$

In view of (82), it then follows that

(89) $$p_i^2 = -2\psi_i(q_i) + \sum_j \Phi_{ij}(q_i)\alpha_j$$
$$\equiv F_i(q_i, \alpha).$$

Thus the phase-path is decomposed into n two-dimensional curves of the form

(90) $$p_i = \pm \sqrt{(F_i(q_i, \alpha))}.$$

If $F \geq 0$, the motion is real and can be classified as *circulation* or *libration*.

A coordinate q_i is said to circulate if it is an angle and if F_i is bounded from above and below by *positive* constants. Depending on the initial conditions, the phase-path is one of the two branches of (90). Since G is a diagonal matrix, (25.1) reduces to

(91) $$p_i = g_{ii}\dot{q}_i$$

with $g_{ii} > 0$. Therefore, if $\dot{q}_i(0) > 0$ then $\dot{q}_i(t) > 0$ and $p_i(t) > 0$ for all t, so that q_i increases monotonically from $q(0)$ to ∞. If $F_i(q)$ is a periodic function, then $p(q)$ is also periodic with a period 2π. As q_i

increases by 2π, the coordinate is said to go through a *cycle*.

A coordinate q_i is said to librate if there exist constants a and b such that

(92)
$$F_i > 0 \quad \text{if } a < q_i < b,$$
$$F_i = 0 \quad \text{if } q_i = a \text{ or } q_i = b, \quad a \leqq q_i(0) \leqq b.$$

It then follows from (91) and the continuity of q_i that $q_i(t)$ is "trapped" in the interval (a, b), traversing it back and forth while the phase-point describes a closed loop $p_i(q_i)$, symmetric about the q-axis. Each closed loop constitutes a cycle of the coordinate.

A simple pendulum librates or circulates, according as $E < 2mgl$ or $E > 2mgl$. If $E = 2mgl$, the phase-path is traversed in infinite time, and the orbit is said to be *asymptotic*.

The *phase-integrals* of the motion are defined by

(93)
$$J_i \equiv \oint p_i(q_i, \alpha) dq_i, \qquad (i = 1, 2, \cdots, n),$$

where the integral is taken over a cycle. Clearly, the n-vector J is a function of the n-vector α only; i.e.,

(94)
$$J = J(\alpha)$$

and is therefore a constant of the motion. Let the Principal Function, written as $W(q, J)$, be the generator of the canonical transformation $(q, p) \rightarrow (w, J)$ with $H' = H = \text{const.}$ Then

(95)
$$p = W_q, \qquad w = W_J.$$

Since $\dot{J} = 0$, w is ignorable, and

(96)
$$H = H(J).$$

Indeed, the canonical equations are

(97)
$$\dot{J} = 0, \qquad \dot{w} = H_J,$$

whose solution is of the form

(98)
$$J = \text{const.},$$
$$w = \nu t + \delta,$$

where ν and δ are constants. The conjugate variables J and w are called the *action-variables* and the *angle-variables* respectively, and

ν is the set of the *fundamental frequencies* of the motion. If (71) is now written in the form

$$(99) \qquad\qquad S = W(q, J) - Ht,$$

and differentiated with respect to J, it then follows with the aid of (95) and (98) that

$$(100) \qquad\qquad \nu = H_J, \qquad \delta = S_J.$$

Thus the fundamental frequencies of the motion are equal to the partial derivatives of the Hamiltonian with respect to the action-variables. The result (98) is a special case of the following theorem.

THEOREM 14. *If all the coordinates are ignorable then the conjugate momenta are constants of the motion and the coordinates are linear functions of the time. The theorem is valid if the words "coordinates" and "momenta" are interchanged.*

We shall next examine the relation between q and w. A function $f(x)$ is said to be *quasi-periodic* if it is a sum of a periodic and a linear function of x; i.e.,

$$(101) \qquad \begin{aligned} f(x) &= rx + g(x), \\ g(x + P) &= g(x). \end{aligned}$$

Such a function is characterized by its *secular rate r* and its *angular frequency $\omega = 2\pi/P$*.

THEOREM 15. *Each (librating/circulating) coordinate q is a (periodic/quasi-periodic) function of the angle variables w. The period in w_i equals unity, and the secular rate equals 2π.*

PROOF. First consider the case where all the coordinates librate. By (95), both w and p are functions of q and the constants J. Hence

$$(102) \qquad dw = w_q dq = W_{Jq} dq = W_{qJ} dq = (\overline{p}\, dq)_J.$$

Let the number of cycles traversed by q lie between m and $m + 1$, where m is a n-vector composed of integers. In view of (93), the integration of (102) over m cycles contributes to Δw the expression

$$(103) \qquad\qquad (\overline{m}J)_J = m;$$

the entire Δw is therefore

(104) $$\Delta w = m + \int_{q(0)}^{q(t)} \bar{p}_J \, dq < m + 1,$$

where the integral is taken over a "fraction" of a cycle. If q goes *exactly* through m cycles, then $q(t) = q(0)$, and

(105) $$\Delta w = m,$$

and conversely. If a coordinate circulates, then it advances by 2π each cycle. In the analysis, $q(0)$ is replaced by $2\pi m$, without affecting the conclusion (105). Therefore q_i can be expressed in terms of the w's in the form

(106) $q_i = 2\pi w_i c + \sum A_j^{(i)} \exp\left[2\pi \sqrt{(-1)} (j_1 w_1 + j_2 w_2 + \cdots) \right],$

where $j_1, j_2, \cdots, j_n$ are integers, and c assumes the value 0 if q_i librates, and the value 1 if it circulates.

In view of (98) two types of motion are distinguished:

(1) If ν_i are commensurable, then $q(t)$ is a periodic function of the time, and the orbit is said to be *periodic*.

(2) If ν_i are incommensurable, then $q(t)$ is not a periodic function of the time, and the orbit is said to be *conditionally-periodic*.

It is to be noted that (105) cannot be satisfied in the actual motion if ν_i are incommensurable. However, this circumstance does not affect the proof of Theorem 15, which deals only with the mathematical properties of the function $q(w)$.

THEOREM 16. *In a conditionally-periodic system the mean frequency of the coordinate q_i averaged over an infinite number of cycles is equal to the fundamental frequency ν_i of the motion.*

PROOF. The set defined by

(107) $$\xi_i = \nu_i / \nu_1 \qquad (i = 1, 2, \cdots, n)$$

contains at least one irrational number. By a theorem of Dirichlet, the system of inequalities

(108) $$|\xi_i - m_i / N| < N^{-1-1/n}$$

has an infinite number of integer solutions for N and m_i. The division of (104) by the time interval T defined by $\nu_1 T = N$ yields with the aid of (98)

(109)
$$\left| \int_{q(0)}^{q(t)} p_J \cdot dq \right| = |N\xi_i - m_i| < N^{-1/n},$$

and

(110)
$$\nu - \frac{m}{T} < \nu_1 N^{-1-1/n}.$$

Therefore the mean frequency ν_* is given by

(111)
$$\nu_* = \lim_{T \to \infty} m/T = \nu.$$

We shall redefine J and w by means of

$$J_i = \frac{1}{2\pi} \oint p_i \, dq_i,$$

(112)
$$w_i = 2\pi(\nu_i t + \delta_i),$$

$$= n_i t + \sigma_i.$$

That the new variables are canonically conjugate follows from Corollary 2 of Theorem 7. The new $w_i(t)$ are angles described with the *angular frequencies* n_i; the initial values $w_i(0) = \sigma_i$ are the *phase-constants*.

VIII. **Application to the Kepler problem.** The Kepler two-body problem is a simple example of a separable system. In the spherical coordinate system (r, θ, ϕ) with θ the complement of the polar angle, for a particle of unit mass,

(113)
$$V = -\mu/r,$$

$$T = \frac{1}{2}(\dot{r}^2 + r^2\dot{\theta}^2 + r^2\cos^2\theta\dot{\phi}^2),$$

where μ is the product of the gravitational constant and the mass of the central body. From (17) and (20),

(114)
$$p_1 = \dot{r}, \quad p_2 = r^2\dot{\theta}, \quad p_3 = r^2\cos^2\theta\dot{\phi}.$$

By Theorem 1,

(115)
$$H = \frac{1}{2}(p_1^2 + p_2^2/r^2 + p_3^2/r^2\cos^2\theta) - \mu/r;$$

the Hamilton-Jacobi equation is

(116) $\qquad (S_r^2 + S_\theta^2/r^2 + S_\phi^2/r^2\cos^2\theta)/2 - \mu/r + S_t = 0.$

Since the Staekel Condition is satisfied in virtue of (86), let

(117) $\qquad S = S_0(t) + S_1(r) + S_2(\theta) + S_3(\phi).$

Then,

(118) $\qquad \begin{aligned} S_0' &= -\alpha_1, \\ S_1' &= (2\alpha_1 + 2\mu/r - \alpha^2/r^2)^{1/2} = p_1, \\ S_2' &= (\alpha_2^2 - \alpha_3^2\sec^2\theta)^{1/2} = p_2, \\ S_3' &= \alpha_3 = p_3, \end{aligned}$

where the α's are the separation constants. (The constants α_2 and α_3 of (82) have been replaced here by $\frac{1}{2}\alpha_2^2$ and $\frac{1}{2}\alpha_3^2$.) The integration of (118) yields

(119) $\qquad S = -\alpha_1 t + \alpha_3\phi + \int p_1\,dr + \int p_2\,d\theta.$

From Theorem 11 follow the kinematical equations,

(120) $\qquad \begin{aligned} \int dr/p_1 &= \beta_1 + t, \\[2mm] -\alpha_2\int dr/r^2 p_1 + \alpha_2\int d\theta/p_2 &= \beta_2, \\[2mm] \phi - \alpha_3\int \sec^2\theta\,d\theta/p_2 &= \beta_3. \end{aligned}$

For a bounded orbit, $\alpha_1 < 0$, the integrals are evaluated as follows: Define new constants $a, e, i, \sigma, \omega, \Omega;\ n$ by the equations

(121) $\qquad \begin{aligned} \alpha_1 &= -\mu/2a, & \alpha_2^2 &= \mu a(1 - e^2) \equiv \mu p, & \cos i &= \alpha_3/\alpha_2, \\ \beta_1 &= \sigma/n, & \beta_2 &= \omega, & \beta_3 &= \Omega, \\ & & a^3 n^2 &= \mu, \end{aligned}$

and the *uniformizing variables* E, v, l, ψ by

(122) $\qquad \begin{aligned} r &= a(1 - e\cos E), \\ r &= p/(1 + e\cos v), \\ l &= E - e\sin E, \\ \sin\theta &= \sin i\,\sin\psi. \end{aligned}$

Then,

(123) $dr = ae \sin E\, dE,$ $-d(1/r) = e \sin v\, dv/p,$

and

$$I_1 \equiv \int_{\min r}^{r} dr/p_1$$

(124) $$= \int r(\mu a)^{-1/2} \left[\, (1 + e - r/a)(r/a - 1 + e)\,\right]^{-1/2} dr$$

$$= \frac{1}{n}(E - e \sin E) = l/n,$$

$$I_2 \equiv \int_{\min r}^{r} \alpha_2\, dr/r^2 p_1$$

(125) $$= -\alpha_2 \int \{(\mu/a)\,[a(1+e)/r - 1][1 - a(1-e)/r]\}^{-1/2} d(1/r)$$

$$= v,$$

$$I_3 \equiv \int_0^{\theta} \alpha_2\, d\theta/p_2 = \int_0^{\theta} \alpha_2(\alpha_2^2 - \alpha_3^2 \sec^2 \theta)^{-1/2} d\theta$$

(126) $$= \int_{/0}^{\theta} (\sin^2 i - \sin^2 \theta)^{-1/2} d(\sin \theta)$$

$$= \psi,$$

(127)
$$I_4 \equiv \int_0^{\theta} \alpha_3 \sec^2 \theta(\alpha_2^2 - \alpha_3^2 \sec^2 \theta)^{-1/2} d\theta = \cos i \int_0^{\psi} \sec^2 \theta\, d\psi$$

$$= \cos i \int d\psi/(1 - \sin^2 i \sin^2 \psi) = \tan^{-1}(\cos i \tan \psi).$$

The equations of the orbit now appear in the form:

$$E - e \sin E = nt + \sigma, \qquad r = a(1 - e \cos E),$$

(128)
$$\tan \frac{v}{2} - = \sqrt{\left(\frac{1+e}{1-e}\right)} \tan \frac{E}{2},$$

$$\psi = v + \omega, \qquad \sin \theta = \sin i \sin \psi,$$

$$\phi = \Omega + \tan^{-1}(\cos i \tan \psi).$$

It is easy to show that (1) the orbit is a plane curve; (2) the plane of the orbit is specified by its *inclination* i and the *longitude of the ascending* node Ω; (3) v is the angle made by the radius vector with a fixed line in the plane; (4) hence the orbit is an ellipse, the fixed line is the line of *apses*, ω is the *argument* of the *pericenter,* and v is the *true anomaly*; (5) E is the *eccentric anomaly, l* the *mean anomaly, n* the *mean motion*; (6) ψ is the argument of latitude.

IX. **The orbital elements.** The elements of a Kepler orbit are the six quantities by which the orbit can be specified. We shall discuss six types of such elements.

1. The Jacobi elements are the constants α and β introduced in the solution of Hamilton-Jacobi equation:

α_1—the energy

α_2—the angular momentum

α_3—the axial component of the angular momentum

β_1—minus the time of the *pericenter passage*

β_2—the argument of the pericenter

β_3—the longitude of the ascending node

2. The Kepler elements are the constants $a, e, i, \sigma, \omega, \Omega$ introduced in §VIII:

$a = -\mu/2\alpha_1$, the semi-axis;

$e = [1 + 2\alpha_1\alpha_2^2/\mu^2]^{1/2}$, the eccentricity;

$i = \cos^{-1}(\alpha_3/\alpha_2)$, the inclination;

$\sigma = \beta_1 n = (-2\alpha_1)^{3/2}\beta_1/\mu$, the mean anomaly at $t = 0$;

$\omega = \beta_2$, the argument of the pericenter;

$\Omega = \beta_3$, the longitude of the ascending node.

3. The action and angle variables are the J's and the w's defined in (112). In terms of the α's, we calculate:

$$J_1 = \frac{1}{2\pi} \oint (2\alpha_1 + 2\mu/r - \alpha_2^2/r^2)^{1/2} dr = \mu(-2\alpha_1)^{-1/2} - \alpha_2,$$

$$(129) \quad J_2 = \frac{1}{2\pi} \oint (\alpha_2^2 - \alpha_3^2 \sec^2\theta)^{1/2} d\theta = \alpha_2 - \alpha_3,$$

$$J_3 = \frac{1}{2\pi} \oint_0^{2\pi} \alpha_3 d\phi = \alpha_3.$$

The integration in the first two lines of (129) has been carried out by means of partial differentiation with respect to the α's. Indeed, with the aid of (124-127),

$$J_{1\alpha_1} = \frac{1}{2\pi} I_1 \Big|_{E=0}^{E=2\pi} = \frac{1}{n} = \mu(-2\alpha_1)^{-3/2},$$

$$J_{1\alpha_2} = -\frac{1}{2\pi} I_2 \Big|_{v=0}^{v=2\pi} = -1,$$

(130)

$$J_{2\alpha_2} = \frac{1}{2\pi} I_3 \Big|_{v=0}^{v=2\pi} = 1,$$

$$J_{2\alpha_3} = -\frac{1}{2\pi} I_4 \Big|_{\psi=0}^{\psi=2\pi} = -1.$$

From the relation $dJ = J_\alpha d\alpha$ we derive $J = \int J_\alpha d\alpha + c$, and (129) follows immediately within the additive constant c. To show that $c = 0$, we note that (1) $e = 0$ implies $r \equiv 0$, $p_1 \equiv 0$, $J_1 = 0$, $2\alpha_1 \alpha_2^2 + \mu^2 = 0$; (2) $i = 0$ implies $\theta \equiv 0$, $p_2 \equiv 0$, $J_2 = 0$, $\alpha_2 = \alpha_3$.

The α's can be expressed in terms of the J's:

$$\sum_1^3 J_i = \mu(-2\alpha_1)^{-1/2} = \sqrt{(\mu a)},$$

$$\alpha_1 = -\frac{1}{2}\mu^2(J_1 + J_2 + J_3)^{-2},$$

(131)

$$\alpha_2 = J_2 + J_3,$$

$$\alpha_3 = J_3.$$

In view of (100), (112), and the relation $H = \alpha_1$, we calculate the fundamental angular frequencies n_i and the phase-constants σ_i from the equations

(132) $\qquad n_i = \alpha_{1J_i}, \quad \sigma_i = S_{J_i}, \qquad (i = 1, 2, 3).$

The first equation yields

(133) $\quad n_1 = n_2 = n_3 = \mu^2(J_1 + J_2 + J_3)^{-3} = \mu^{-1}(-2\alpha_1)^{3/2} = n.$

Since the frequencies are commensurable, the motion is periodic; since they are equal, the motion is said to be *degenerate*. In view of identities (1) and (3) of §II, the second equation of (132) can be written in the form

(134) $\qquad\qquad\qquad \sigma = S_J = \overline{\alpha}_J S_\alpha.$

By the differentiation of the α's in (131) we obtain

$$(135) \qquad \overline{\alpha}_J = \begin{pmatrix} n & 0 & 0 \\ n & 1 & 0 \\ n & 1 & 1 \end{pmatrix}.$$

Since $S_\alpha = \beta$, (134) then leads to

$$\begin{aligned} \sigma_1 &= n\beta_1 = \sigma, \\ (136) \qquad \sigma_2 &= n\beta_1 + \beta_2 = \sigma + \omega, \\ \sigma_3 &= n\beta_1 + \beta_2 + \beta_3 = \sigma + \omega + \Omega. \end{aligned}$$

Finally, the angle-variables w_i are constructed with the aid of (112):

$$\begin{aligned} w_1 &= nt + \sigma = l, \\ (137) \qquad w_2 &= l + \omega, \\ w_3 &= l + \omega + \Omega. \end{aligned}$$

4. The canonical elements of Delaunay can be obtained from the action and angle-variables (w, J) by means of a linear extended-point transformation. The latter will be generated by a function S of the form

$$(138) \qquad S = \overline{q}' A p,$$

where A is a constant matrix. By Theorem 11, the transformation defined by

$$p' = Ap, \qquad q = \overline{A}q'$$

is canonical. As a check, we note that the relations

$$(139) \qquad p_p' = A, \qquad q_q' = \overline{A}^{-1}$$

indeed satisfy the hypothesis of Theorem 7. Let $q = w$, $p = J$, and choose A as

$$(140) \qquad A = \begin{pmatrix} 1 & 1 & 1 \\ 0 & 1 & 1 \\ 0 & 0 & 1 \end{pmatrix}.$$

Then, the contragredient of A is

(141)
$$\overline{A}^{-1} = \begin{pmatrix} 1 & 0 & 0 \\ -1 & 1 & 0 \\ 0 & -1 & 1 \end{pmatrix},$$

and the new variables are constructed from

(142)
$$q' = \overline{A}_q^{-1}, \qquad p' = Ap,$$

which yields

(143)
$$q' = \begin{pmatrix} l \\ \beta_2 \\ \beta_3 \end{pmatrix}, \qquad p' = \begin{pmatrix} \sqrt{(\mu a)} \\ \alpha_2 \\ \alpha_3 \end{pmatrix}.$$

Delaunay adopted the notation $L = \sqrt{(\mu a)}$, $G = \alpha_2$, $H = \alpha_3$, $g = \beta_2$, $h = \beta_3$. In terms of the Kepler elements,

(144)
$$\begin{aligned} L &= \sqrt{(\mu a)}, & l &= nt + \sigma, \\ G &= L\sqrt{(1 - e^2)}, & g &= \omega, \\ H &= G\cos i, & h &= \Omega. \end{aligned}$$

Conversely, the relations

(145)
$$\begin{aligned} w_1 &= l, & J_1 &= L - G, \\ w_2 &= l + g, & J_2 &= G - H, \\ w_3 &= l + g + h, & J_3 &= H, \end{aligned}$$

express the action and angle-variables in terms of the elements of Delaunay.

Since the letter H has been pre-empted by Delaunay, it has become customary in celestial mechanics to use the letter F for the Hamiltonian. Another convention is the change of sign of the Hamiltonian, which is equivalent to the interchange of the coordinates and momenta. Thus

(146)
$$\begin{aligned} F &= -\alpha_1 = \mu^2/2L^2, \\ \dot{L} &= F_l = 0, \qquad \dot{l} = -F_L = \mu^2/L^3 = n. \end{aligned}$$

With the exception of l, the Delaunay elements are constants in the undisturbed motion.

An alternate derivation of Delaunay elements is based on the generating function

(147) $S = -\alpha_1 t + \mu(-2\alpha_1)^{-1/2}\beta_2' + \alpha_2\beta_{1'}^+ + \alpha_3\beta_3',$

where the unprimed α and β are the previously defined elements of Jacobi. Then $\alpha_1' = S_{\beta_1'}$ and $\beta_1 = S_{\alpha_1}$, leading to

$$\alpha_1' = \mu(-2\alpha_1)^{-1/2} = \sqrt{(\mu a)} \equiv L,$$

(148) $\beta_1 = -t + \mu(-2\alpha_1)^{-3/2}\beta_1' = -t + \beta_1'/n,$

$$\beta_1' = n(t + \beta_1) = nt + \sigma = l.$$

The remaining elements are unchanged. Since the old Hamiltonian corresponding to the Jacobi elements is zero, the new Hamiltonian becomes

(149) $F' = F + S_t = -\alpha_1 = \mu^2/2L^2,$

in agreement with (146).

The "slow" Delaunay set, obtained by the replacement of l by the constant σ, can be derived from S of (147) with the term $-\alpha_1 t$ omitted. The corresponding Hamiltonian is zero in the undisturbed motion, in contrast to (150), which belongs to the original, or the "fast" set.

5. A modified Delaunay set is constructed by a canonical linear transformation of the original set. With S of the form (138), and

(150) $A = \begin{pmatrix} 1 & 0 & 0 \\ -1 & 1 & 0 \\ 0 & -1 & 1 \end{pmatrix},$

we first construct the contragredient matrix,

(151) $\bar{A}^{-1} = \begin{pmatrix} 1 & 1 & 1 \\ 0 & 1 & 1 \\ 0 & 0 & 1 \end{pmatrix},$

and then derive the new variables from (139) as

(152) $L, G - L, H - G; \quad l + g + h, \quad g + h, h.$

We recognize $l + g + h$ as the *mean longitude*, and $g + h$ as the *longitude of the pericenter*. It is noteworthy that

$$G - L = O(e^2),$$

(153)

$$H - G = O(\sin^2 i/2).$$

6. The Poincaré elements,

$$L, \qquad\qquad\qquad l + g + h,$$
$$(154) \quad \sqrt{(2(L - G))}\sin(g + h), \quad \sqrt{(2(L - G))}\cos(g + h),$$
$$\sqrt{(2(G - H))}\sin h, \qquad \sqrt{(2(G - H))}\cos h,$$

are derived from the modified Delaunay elements by means of the transformation

$$(155) \quad q_i' = \sqrt{(-2q_i)}\sin p_i, \quad p_i' = \sqrt{(-2q_i)}\cos p_i \quad (i = 2, 3).$$

That this transformation is canonical follows from the fact that the Jacobian determinant is $+1$ in each of the three phase subspaces. Hence the Jacobi relation (45.2) is satisfied.

Poincaré elements are especially useful for orbits of small eccentricity and inclination. Indeed,

$$(156) \quad \begin{aligned} q_2' &\sim e\sin\varpi, & p_2' &\sim e\cos\varpi, \\ q_3' &\sim \sin(i/2)\sin h, & p_3' &\sim \sin(i/2)\cos h, \end{aligned}$$

where $\varpi \equiv g + h$.

X. The method of variation of constants. A nonseparable system with a small parameter k can be solved approximately. In planetary theory, k is the ratio of the mass of the disturbing planet to the mass of the Sun; in the lunar theory it is the ratio of the mean motion of the Sun to that of the Moon; in the artificial satellite theory k is J_2, the coefficient of the second zonal harmonic of the geopotential. In all cases, the solution can be expressed as a power series in k.

The method of variation of constants involves the following four steps:

1. Choose a "reference" potential V_0 for which the Staekel Condition is satisfied and for which the "disturbing" potential $V_1 \equiv V - V_0$ is of order k.

2. Solve the Hamilton-Jacobi equation

$$(157) \qquad\qquad H_0(q, S_q, t) + S_t = 0,$$

corresponding to $V = V_0$, yielding $S(q, \alpha; t)$; then $S_\alpha = \beta$ furnishes the solution $q^{(0)}(t; \alpha, \beta)$ of the "unperturbed" problem. (Examples: the Kepler ellipse for $V_0 = -1/r$ and the Garfinkel and the Vinti "intermediaries" in the artificial satellite theory.)

3. Let $H_1 \equiv H - H_0 = O(k)$, where H is the Hamiltonian of the original problem, and let S above be the generating function of the canonical transformation $(q, p) \to (\beta, \alpha)$. The new Hamiltonian is

$$(158) \qquad H' = H + S_t = H_0 + S_t + H_1 = H_1,$$

and the new variables (β, α) satisfy

$$(159) \qquad \dot{\beta} = H_{1\alpha}, \qquad \dot{\alpha} = - H_{1\beta}.$$

In matrix form,

$$(160) \qquad \dot{e} = \Phi_0 H_{1e}, \qquad e \equiv \begin{pmatrix} \beta \\ \alpha \end{pmatrix}.$$

These are the equations of the variation of the elements.

4. Solve the equations of variation by successive approximation, such as the method of Poisson and the method of Delaunay. An epitome of the entire procedure was given by T. E. Sterne: "In order to solve the exact problem approximately, we first solve an approximate problem exactly."

An independent derivation of (160) is contained in the following theorem.

THEOREM 17. *If the solution of the problem with $H = H_0 + H_1$ is expressed in the same form $x = x_{(0)}(t, e)$ as that of the unperturbed problem with $H = H_0$, but with e no longer constant, then e satisfies*

$$(161) \qquad \dot{e} = \Phi H_{1e},$$

where

$$(162) \qquad \Phi \equiv e_x \Phi_0 \overline{e}_x.$$

PROOF. From (36),

$$(163) \qquad \dot{x} = \Phi_0 H_x, \qquad \dot{x}_{(0)} = \Phi_0 H_{0x}.$$

By the differentiation of $x = x(t, e)$,

$$(164) \qquad \dot{x} = x_t + x_e \dot{e}.$$

Since $x = x_{(0)}$ corresponds to $e = \text{const.}$,

$$(165) \qquad \dot{x}_{(0)} = x_t.$$

Let D_x and D_e denote the gradient operators in the x-space and the e-space respectively. Then

(166) $$x_e \hat{e} = \Phi_0(H - H_0)_x = \Phi_0 H_{1x} = \Phi_0 D_x H_1.$$

By properties (6) and (7) of §II $(x_e)^{-1} = e_x$ and $D_x = \overline{e}_x D_e$. The conclusion now follows with the aid of (162).

If e is a canonical set of constants, such as

$$e = \begin{pmatrix} \beta \\ \alpha \end{pmatrix},$$

then $\Phi = \Phi_0$. In accord with the convention of §IX the coordinates and the momenta are interchanged while the sign of the Hamiltonian is reversed. If the *disturbing function R* is defined by $R = -H_1$, then the canonical equations for the variation of constants take the form

(167) $$\dot{\alpha} = R_\beta, \qquad \dot{\beta} = -R_\alpha,$$

or

(168) $$\dot{e} = \Phi_0 R_e, \qquad e \equiv \begin{pmatrix} \alpha \\ \beta \end{pmatrix}.$$

XI. **On the Poisson brackets.** The equations of variation for a non-canonical set of variables involve the Poisson matrix Φ. The calculation of Φ is facilitated by the use of the following two theorems.

THEOREM 18. *The Poisson matrix $\Phi(e)$ belonging to a set of variables e is a function of e only, and is independent of the reference canonical variables x appearing in the definition* (162).

PROOF. Let x and y be two sets of canonical variables, and let $\Phi(e \mid x)$ denote $\Phi(e)$ defined with respect to x. For the canonical transformation $y \to x$, with $J = x_y$, (162) and (44) imply

(169) $$\Phi(e \mid x) = e_x \Phi_0 \overline{e}_x = e_x(x_y \Phi_0 \overline{x}_y)\overline{e}_x.$$

By the *associative* property of matrix multiplication, property (4) of §II, and the *reversal* rule for *transposition*, we obtain

(170) $$\Phi(e \mid x) = (e_x x_y)\Phi_0(\overline{x}_y \overline{e}_x) = e_y \Phi_0 \overline{e}_y = \Phi(e \mid y).$$

THEOREM 19. *If the two 2n-vectors*

$$e' = \begin{pmatrix} q' \\ p' \end{pmatrix}, \qquad e = \begin{pmatrix} q \\ p \end{pmatrix},$$

have the following properties: (1) *e is a canonical set*; (2) *e′ is derived from e by the semi-identical transformation of the form*

(171) $$q' = q'(q), \qquad p' = p,$$

then the Poisson matrix $\Phi(e')$ *has the structure*

(172) $$\Phi = \begin{pmatrix} 0 & Q \\ -\bar{Q} & 0 \end{pmatrix},$$

where $Q \equiv q'_q$.

The proof follows from (43) with

(173) $$J = e'_e = \begin{pmatrix} Q & 0 \\ 0 & I \end{pmatrix}.$$

As an example, let e' be the Kepler set of elements and e the "slow" Delaunay set of §IX. The semi-identical transformation $e \to e'$ is characterized by

(174) $$Q = \begin{pmatrix} a_L & 0 & 0 \\ e_L & e_G & 0 \\ 0 & i_G & i_H \end{pmatrix}.$$

The partial derivatives are obtained from

(175) $$a = L^2/\mu, \qquad e = (1 - G^2/L^2)^{1/2}, \qquad \cos i = H/G,$$
$$\mu = n^2 a^3,$$

leading to

(176) $$\begin{aligned} a_L &= 2L/\mu = 2/na \equiv (1, 4), \\ e_L &= (1 - e^2)/na^2 e \equiv (2, 4), \\ e_G &= -(1 - e^2)^{1/2}/na^2 e \equiv (2, 5), \\ i_G &= \cot i/na^2(1 - e^2)^{1/2} \equiv (3, 5), \\ i_H &= -\csc i/na^2(1 - e^2)^{1/2} \equiv (3, 6), \end{aligned}$$

the remaining derivatives being zero.

With e replaced by e', (161) becomes the celebrated equations of Lagrange for the variation of the Kepler elements. Explicitly,

$$\dot{a} = (1,4)\, R_\sigma,$$
$$\dot{e} = (2,4)\, R_\sigma + (2,5)\, R_\omega,$$
$$\dot{i} = (3,5)\, R_\omega + (3,6)\, R_\Omega,$$
(177)
$$\dot{\sigma} = -\,(1,4)\, R_a - (2,4)\, R_e,$$
$$\dot{\omega} = -\,(2,5)\, R_e - (3,5)\, R_i,$$
$$\dot{\Omega} = -\,(3,6)\, R_i.$$

It should be noted that a enters R explicitly through r, as well as implicitly through $l = nt + \sigma$, with $n^2 a^3 = \mu$. Therefore,

(178)
$$\dot{\sigma} = -\,(1,4)\,(R_a) - (1,4)\, n_a R_\sigma t - (2,4)\, R_e,$$
$$\dot{a} = (1,4)\, R_\sigma,$$

where (R_a) denotes explicit differentiation. Since $n_a \dot{a} = \dot{n}$,

(179)
$$\dot{\sigma} + \dot{n}t = -\,(1,4)\,(R_a) - (2,4)\, R_e.$$

The appearance of t outside the trigonometric term can therefore be avoided if we use, instead of σ, the element σ_* defined by

(180)
$$l = \int n\,dt + \sigma_*.$$

Then $\dot{\sigma}_*$ and (R_a) replace $\dot{\sigma}$ and R_a respectively throughout (177). As a result of this transformation, double quadrature is required to calculate the perturbation δl. Indeed,

$$\delta l = \int \delta n\,dt + \delta \sigma_*,$$
(181)
$$\delta n = -\,(3n/2a)\delta a,$$
$$\delta a = \int (2R_\sigma/na)\,dt,$$

and hence

(182)
$$\delta l = -\iint (3R_\sigma/a^2)\,dt\,dt + \delta \sigma_*.$$

Another form of the Lagrange equations uses the modified Kepler elements a, e, i, ϵ, ϖ, Ω, where ϵ is the *mean longitude* at $t = 0$, and ϖ is the longitude of the pericenter,

(183)
$$\epsilon \equiv \sigma_* + \varpi,$$
$$\varpi = \omega + \Omega.$$

The effect of this transformation is to replace Q of (174) by

$$(184) \qquad Q' = Q \begin{pmatrix} 1 & 0 & 0 \\ 1 & 1 & 0 \\ 0 & 1 & 1 \end{pmatrix}.$$

The new Poisson brackets that are affected are:

$$(2,4)' = (2,4) + (2,5) = -(1-e^2)^{1/2}[1-(1-e^2)^{1/2}]/na^2 e,$$

$$(185) \quad (3,4)' = (3,5) + (3,6) = -\tan\frac{1}{2}\,i/na^2(1-e^2)^{1/2},$$

$$(3,5)' = (3,4)',$$

and (177) becomes

$$(186) \qquad \begin{aligned} \dot{a} &= (1,4)\,R_\epsilon, \\ \dot{e} &= (2,4)'\,R_\epsilon + (2,5)\,R_{\varpi}, \\ \dot{i} &= (3,4)'\,(R_\epsilon + R_{\varpi}) + (3,6)\,R_\Omega, \\ \dot{\epsilon} &= -(1,4)\,R_a - (2,4)'\,R_e - (3,4)'\,R_i, \\ \dot{\varpi} &= -(2,5)\,R_e - (3,5)'\,R_i, \\ \dot{\Omega} &= -(3,6)\,R_i. \end{aligned}$$

XII. On the Poisson method. The equations of variation of the elements are of the form

$$(187) \qquad \dot{x} = \Phi DR = kf(x,t); \qquad (k \ll 1).$$

The variables x and f are expanded into a Taylor series about the undisturbed orbit $x = x^{(0)}$ corresponding to $k = 0$. We have

$$(188) \qquad f = f_0 + (f_x)_0\,\delta x + \frac{1}{2}\delta\overline{x}\cdot(f_{xx})_0\cdot\delta x + \cdots,$$

where

$$(189) \qquad \delta x \equiv x - x^{(0)} \equiv kx^{(1)} + k^2 x^{(2)} + \cdots.$$

The two expansions are substituted into (187), and the coefficients of like powers of k are equated, yielding

$$(190) \qquad \begin{aligned} \dot{x}^{(0)} &= 0, \\ \dot{x}^{(1)} &= f_0, \\ \dot{x}^{(2)} &= (f_x)_0\,x^{(1)}, \\ \cdots &\quad \cdots. \end{aligned}$$

Thus we have

$$x^{(0)} = \text{const.}$$

(191)
$$x^{(1)} = \int f(x^{(0)}, t)dt,$$

$$x^{(2)} = \int f_x(x^{(0)}, t)x^{(1)} dt,$$

$$\cdots \qquad \cdots$$

for the perturbations of successive orders.

Consider a problem of two degrees of freedom with R in the form of a Fourier series in the angle variables w with coefficients functions of the action variables J. Then

$$(192) \qquad R = \sum_{0,-\infty}^{\infty} A_{j_1 j_2}(J)\cos(j_1 n_1 + j_2 n_2)t,$$

and x^p is of the form

$$(193) \quad x^p = B_{00}t + \sum B_{j_1 j_2}\exp\left[\sqrt{(-1)}(j_1 n_1 + j_2 n_2)t\right](j_1 n_1 + j_2 n_2)^{-1}.$$

The perturbation terms can be classified as follows:

(1) Secular term $B_{00}t$, $\qquad j_1 = j_2 = 0$,
(2) Short-periodic terms, $\qquad j_1 n_1 + j_2 n_2 = O(kn)$.
(3) Long-periodic terms, $\qquad j_1 n_1 + j_2 n_2 = O(n)$,

We observe that (1) only the angle variables have secular perturbations; (2) in the integration of the long-periodic terms there occurs a reduction of the order by unity; i.e., first-order effects appear in the expression for $x^{(2)}$, etc.

In the artificial satellite theory, the long-periodic terms have the argument $2g = 2(w_2 - w_1)$, which corresponds to $j_1 + j_2 = 0$ with

$$n_2 - n_1 = \frac{3}{4}\mu^2 J_2 n_1(5\cos^2 i - 1),$$

(194)
$$k \equiv J_2 \sim 10^{-3}.$$

At resonance, $i \sim 63°.4$, where the classical solution breaks down, special treatment is required.

XIII. **On the Delaunay method.** Let the Hamiltonian F be of the form

$$(195) \qquad F = \sum_{0,-\infty}^{\infty} A_{ij}(L, G) \cos(il + jg).$$

As in the preceding section, the terms of F are classed as secular $(i = j = 0)$, the long-periodic $(i = 0, j \neq 0)$, and the short-periodic $(i \neq 0)$ terms. Let the corresponding portions of F be denoted by the bar, the star, and the tilde; i.e.

$$\bar{F} = A_{00}, \qquad F^* = \sum_{1}^{\infty} A_{0j} \cos jg,$$

(196)

$$\tilde{F} = F - \bar{F} - F^*.$$

Delaunay used a succession of canonical transformations to remove, one by one, the periodic terms of F. In the limit the new Hamiltonian F' becomes purely secular; i.e. a function of L' and G' only. Then by Theorem 14, L' and G' are constants of the motion, and l', g' are linear functions of the time. With the generating function S correlating the new and the old variables, the problem is solved. The feasibility of a single transformation to accomplish the same purpose was suggested by von Zeipel. It is often convenient to remove first the terms of short period.

XIV. **On the von Zeipel method.** The von Zeipel method (1916) will be illustrated here by a problem of two degrees of freedom. Without any loss of generality, we denote the canonical variables of the system by L, G, l, g, where l is the "fast" variable appearing in the short-periodic terms, and g is the "slow" variable that characterizes the terms of long period. Furthermore, let the undisturbed Hamiltonian F_0 be a function of L only. Then the undisturbed mean motion n is given by

(197) $$n = -F_{0L}.$$

We seek a transformation $(L, G, l, g) \rightarrow (L', G', g', h')$ such that the new Hamiltonian F' is a function of L', G' only. As noted in the previous section, the problem of motion is solved if we can determine F' and the generating function $S(L', G', l, g)$.

We write

(198)
$$F = F_0 + F_1 + F_2 + \cdots,$$
$$F' = F'_0 + F'_1 + F'_2 + \cdots,$$
$$S = L'l + G'g + \tilde{S}_1 + \tilde{S}_2 + S_1^* + \cdots,$$

where the numerical subscripts denote the order of magnitude with respect to the small parameter of the system. The transformation equations are of the type (59.3), and take the form

$$L = S_l = L' + \widetilde{S}_{1l} + \widetilde{S}_{2l} + \cdots,$$

(199)
$$G = S_g = G' + \widetilde{S}_{1g} + S^{*}_{1g} + \cdots,$$

$$l' = S_{L'} = l + \widetilde{S}_{1L'} + S^{*}_{1L'} + \cdots,$$

$$g' = S_{G'} = g + \widetilde{S}_{1G'} + S^{*}_{1G'} + \cdots.$$

The old Hamiltonian F is now expanded about L', G', l, g by means of the Taylor series:

$$F_0 = [F_0 + F_{0L}(\widetilde{S}_{1l} + \widetilde{S}_{2l}) + \frac{1}{2} F_{0LL}\widetilde{S}^2_{1l}] + \cdots,$$

(200)
$$F_1 = [F_1 + F_{1L}\widetilde{S}_{1l} + F_{1G}(\widetilde{S}_{1g} + S^{*}_{1g})] + \cdots,$$

$$F_2 = [F_2] + \cdots.$$

Here the expansion is carried out to the second order, and the bracket is used as the symbol of the substitution $L \to L'$, $G \to G'$. Since $S_t = 0$, (49) yields $F' = F$, and

(201)
$$F'_0 + F'_1 + F'_2 + \cdots =$$
$$[F_0 + F_1 + F_2 + \Phi_2 + F_{1G}S^{*}_{1g} - n(\widetilde{S}_{1l} + \widetilde{S}_{2l})],$$

where Φ_2 is defined by

(202)
$$\Phi_2 \equiv F_{1L}\widetilde{S}_{1l} + F_{1G}\widetilde{S}_{1g} + \frac{1}{2} F_{0LL}\widetilde{S}^2_{1l}.$$

Separation by orders of magnitude converts (201) into

$$F'_0 = [F_0],$$

(203)
$$F'_1 = [F_1 - n\widetilde{S}_{1l}],$$

$$F'_2 = [F_2 + \Phi_2 + F_{1G}S^{*}_{1g} - n\widetilde{S}_{2l}].$$

In our problem, the feasibility of the von Zeipel transformation requires that

(204)
$$F^{*}_1 = 0.$$

If this requirement is satisfied, (203) can be separated by type of term; i.e., the secular, the long, and the short, leading to:

$$F_0' = [F_0],$$
$$F_1' = [\overline{F}_1],$$
$$(205) \qquad 0 = [\tilde{F}_1 - n\tilde{S}_{1l}],$$
$$F_2' = [\overline{F}_2 + \overline{\Phi}_2],$$
$$0 = [F_2^* + \Phi_2^* + \overline{F}_{1G}S_{1g}^*],$$
$$0 = [\tilde{F}_2 + \tilde{\Phi}_2 + \tilde{F}_{1G}S_{1g}^* - n\tilde{S}_{2l}].$$

Now the six equations (205) determine the six unknowns F_0', F_1', F_2', $\tilde{S}_1$, S_1^*, $\tilde{S}_2$. In particular, S is obtained from

$$\tilde{S}_1 = \left[\frac{1}{n} \int \tilde{F}_1 \, dl \right],$$

$$(206) \qquad S_1^* = - \left[\int (F_2^* + \Phi_2^*) dg / \overline{F}_{1G} \right],$$

$$\tilde{S}_2 = \left[\frac{1}{n} \int (\tilde{F}_2 + \tilde{\Phi}_2 + \tilde{F}_{1G}S_{1g}^*) dl \right].$$

From the canonical equations in the new variables it follows that

$$(207) \qquad \begin{aligned} L' &= \text{const.,} & l' &= -(F_{L'}')t + l'(0), \\ G' &= \text{const.,} & g' &= -(F_{G'}')t + g'(0). \end{aligned}$$

Finally, the generalized Kepler equations (199.3) and (199.4) are solved for l and g in terms of the constants L', G' and of the linear functions l', g' of the time.

The appearance of a small divisor $\overline{F}_{1G}$ in the calculation of the long-periodic terms is clearly seen in (206.2). If this divisor vanishes, as in the problem of the Critical Inclination in the Artificial Satellite Theory, it becomes necessary to carry the Taylor expansion to the next higher term.

Generally, the feasibility of the von Zeipel transformation is assured by the following *Poincaré criterion*, which is a generalization of (204): "The order of a periodic term in the Hamiltonian is at least one more than the order of the time derivative of its trigonometric argument."

ACKNOWLEDGMENT. The author expresses his appreciation to Dr. Louis Friedman, who has carefully taken down the lecture notes, and to Mrs. Bernice Krouse and Mrs. Eleanor Fears, who have meticulously typed the manuscript.

References

1. D. Brouwer and G. M. Clemence, *Methods of celestial mechanics,* Academic Press, New York, 1961; Chapters XI, XVII.

2. H. Goldstein, *Classical mechanics,* Addison-Wesley, Reading, Massachusetts, 1950; Chapters 2, 7, 8, 9.

3. C. Lanczos, *Variational principles of mechanics,* Univ. of Toronto Press, Toronto, 1949.

4. E. T. Whittaker, *Analytical dynamics,* Dover, New York, 1944; Chapters X, XI, XII.

5. B. Garfinkel, *On the perturbation matrices of celestial mechanics,* Astronom. J. (2) **51** (1944), 44-48.

6. C. L. Siegel, *Vorlesungen über Himmelsmechanik,* Springer-Verlag, Berlin, 1956.

7. A. Wintner, *Analytical foundations of celestial mechanics,* Princeton Univ. Press, Princeton, N. J., 1947; Chapters I, II, III.

8. T. E. Sterne, Astronom. J. **63** (1958), 28-40.

9. B. Garfinkel, Astronom. J. **63** (1958), 88-96.

10. _____, *The orbit of a satellite of an oblate planet,* Astronom. J. **64** (1959), 353-367.

11. _____, Astronom. J. **69** (1964), 223.

12. J. Vinti, *New method of solution for unretarded satellite orbits,* J. Res. Nat. Bur. Standards Sect. B **63** (1959), 105-116.

13. _____, *Intermediary equatorial orbits of an artificial satellite,* Astronom. J. **66** (1961), 514-516.

14. _____, *Mean motions in conditionally periodic separable systems,* J. Res. Nat. Bur. Standards Sect. B. (2) **65** (1961), 131-135.

BALLISTIC RESEARCH LABORATORIES
ABERDEEN PROVING GROUND, MARYLAND

P. J. Message

Stability and Small Oscillations
about Equilibrium
and Periodic Motions

The general motion in a problem in celestial mechanics may often by exhibited as an oscillation or combination of oscillations about one or another of a number of configurations of absolute or relative equilibrium, or about a periodic motion (which may be one of a series of periodic motions). The periodic motion may be regarded as a generalisation of relative equilibrium, and variables may of course always be found in terms of which it appears as a relative equilibrium configuration. The value of periodic motions as intermediate orbits is illustrated in Hill's Lunar Theory [1] de Sitter's theory of the great satellites of Jupiter [2], and in many theories of the motion of minor planets and satellites involving commensurable or nearly commensurable periods (for example the Trojan group).[1]

Poincaré's treatment of periodic motions in general, and his application of it to the general motion problem of three bodies, leading to the proof of the existence of periodic solutions of three sorts in that problem, was the subject of lectures at the July, 1960, Summer Institute.[2] The first part of the present chapter deals with the

---

[1] See e.g. E. W. Brown, Astronom. J. **35**(1923), 69. Yale Trans. **3**(1923), Parts 1 and 3; D. Brouwer, Yale Trans. **6** (1933), Part 7; W. M. Smart, Mem. Roy. Astronom. Soc. **62**(1918), 79; and H. Hertz, Astronom. J. **50** (1946), 121.

[2] See §§1, 2 and 3 of the notes on Periodic orbits for the July 1960 Summer Institute in Dynamical Astronomy at Yale University, McDonnell Aircraft Corp., St. Louis, Missouri, 1961.

application of Poincaré's procedure to the main problem of the Lunar Theory, and the second part to an investigation of motion in the vicinity of a periodic solution or equilibrium configuration, using the equations of linear variation.

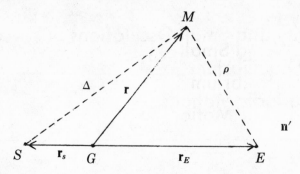

FIGURE 1. Coordinates for Sun, Earth, Moon System

I. Application of Poincaré's procedure to the lunar theory.

a. *Description of problem.* The "Main Problem" of the moon's motion is an idealisation in which we suppose the Sun, S, and the Earth, E, to revolve about their common centre of mass G with angular velocity $\mathbf{n}'$, and suppose the mass of the Moon, M to be so small that its effect on the motion of the other two bodies is negligible. Then (see Figure 1) if $\mathbf{r}$, $\mathbf{r}_S$ and $\mathbf{r}_E$ are the position vectors of M, S and E respectively, referred to G, in a frame which is rotating with angular velocity $\mathbf{n}'$, the equation of motion of the Moon may be written

$$(\text{I.1}) \quad \ddot{\mathbf{r}} + 2\mathbf{n}' \times \dot{\mathbf{r}} + \mathbf{n}' \times (\mathbf{n}' \times \mathbf{r}) = -\frac{Gm_E(\mathbf{r} - \mathbf{r}_E)}{\rho^3} - \frac{Gm_S(\mathbf{r} - \mathbf{r}_S)}{\Delta^3}$$

where m_S, m_E are the masses of the Sun and Earth respectively, G is the constant of gravitation, and

$$\rho = |\mathbf{r} - \mathbf{r}_E| = EM,$$
$$\Delta = |\mathbf{r} - \mathbf{r}_S| = SM.$$

We now use the position vector of the Moon relative to the Earth, $\boldsymbol{\rho} = \mathbf{r} - \mathbf{r}_E$. At the same time put $\mu = Gm_E$ and note that if a' is the radius of the Earth's orbit about the Sun (SE), then

$$Gm_S = n'^2 a'^3$$

very closely. Equation (I.1) then becomes

$$(I.2)\quad \ddot{\rho} + 2\mathbf{n}' \times \dot{\rho} + \mathbf{n}'(\mathbf{n}' \cdot \rho) - n'^2(\mathbf{r}_E + \rho) = -\frac{\mu\rho}{\rho^3} - \frac{n'^2 a'^3(\rho + \mathbf{a}')}{\Delta^3},$$

where $\mathbf{a}' = \mathbf{r}_E - \mathbf{r}_S = \mathbf{SE} = \mathbf{r}_E$ very closely. Now

$$\Delta^2 = (\mathbf{a}' + \rho)^2$$
$$= a'^2 + 2\mathbf{a}' \cdot \rho + \rho^2,$$

so

$$\frac{1}{\Delta^3} = \frac{1}{a'^3} - \frac{3\mathbf{a}' \cdot \rho}{a'^5} + \frac{1}{a'^3} O\left(\frac{\rho^2}{a'^2}\right),$$

and so equation (I.2) becomes, collecting terms, writing $\mathbf{a}'$ for $\mathbf{r}_E$, and neglecting ρ^2 in comparison with a'^2,

$$(I.3)\qquad \ddot{\rho} + 2\mathbf{n}' \times \dot{\rho} + \mathbf{n}'(\mathbf{n}' \cdot \rho) = -\frac{\mu\rho}{\rho^3} + \frac{3n'^2(\mathbf{a}' + \rho)}{a'}.$$

(The latter approximation is equivalent to taking the limit as the Sun's distance (a') tends to infinity while the mass also tends to infinity in such a way that the angular speed (n') of the Earth about it remains constant.) Taking a system of axes with E as origin, axis of ξ in the direction of $\mathbf{SE}$, axis of η in the plane of the Earth's motion, and axis of ζ completing the right handed orthogonal set, we have

$$\mathbf{n}' = (0, 0, n'), \quad \mathbf{a}' = (a', 0, 0), \quad \rho = (\xi, \eta, \zeta),$$

and so the equation (I.3) becomes

$$\ddot{\xi} - 2n'\dot{\eta} - 3n'^2\xi = -\frac{\mu\xi}{\rho^3},$$

$$(I.4)\qquad\qquad \ddot{\eta} + 2n'\dot{\xi} = -\frac{\mu\eta}{\rho^3},$$

$$\ddot{\zeta} + n'^2\zeta = -\frac{\mu\zeta}{\rho^3}.$$

b. *The circular solution, and motion in its vicinity.* In the absence of the Sun, the last term of (I.2) does not appear, and also G coincides with E, so that $\mathbf{r}_E = \mathbf{O}$. The equations are then

$$\ddot{\xi} - 2n'\dot{\eta} - n'^2\xi = -\frac{\mu\xi}{\rho^3},$$

(I.5) $$\ddot{\eta} + 2n'\dot{\xi} - n'^2\eta = -\frac{\mu\eta}{\rho^3},$$

$$\ddot{\zeta} + n'^2\zeta = -\frac{\mu\zeta}{\rho^3}$$

which have the periodic solution

(I.6) $\xi = a\cos\nu t,$ $\eta = a\sin\nu t,$ $\zeta = 0$

where a is a constant and $(\nu + n')^2 a^3 = \mu$. This represents an undisturbed circular motion, in the plane of the Earth's orbit. We will study possible motions in the vicinity of this circular orbit, in its plane, in the presence of the Sun under the conditions of equations (I.3) and (I.4). So we put

(I.7) $\xi = a\cos\nu t + \delta\xi,$ $\eta = a\sin\nu t + \delta\eta,$ $\zeta = 0.$

Then $\rho^2 = \xi^2 + \eta^2 = a^2 + 2a(\cos\nu t\delta\xi + \sin\nu t\delta\eta) + O(\delta\xi^2, \delta\eta^2)$ and equations (I.4) become, putting (with $i = (-1)^{1/2}$)

$$w = \delta\xi(xi) + i\delta\eta,   z = \exp i\nu t,$$

(I.8) $$\ddot{w} + 2in'\dot{w} - n'^2 w - \frac{1}{2}(n' + \nu)^2(w + 3z^2\overline{w})$$

$$= \frac{1}{2}n'^2 a(z + 3z^{-1}) + O(w^2).$$

In order to seek a complementary function, we try

$$w = Az^\alpha + Bz^{2-\alpha},$$

where A, B and α are constants to be determined. Substitution into (I.8) leads to

$$-\left\{(\alpha\nu + n')^2 + \frac{1}{2}(\nu + n')^2\right\}Az^\alpha - \frac{3}{2}(\nu + n')^2\overline{A}z^{2-\alpha}$$

$$-\left\{(2\nu - \alpha\nu + n')^2 + \frac{1}{2}(\nu + n')^2\right\}Bz^{2-\alpha}$$

$$-\frac{3}{2}(\nu + n')^2\overline{B}z^\alpha = 0.$$

The coefficients of z^α and $z^{2-\alpha}$ must each vanish, and if A and B are not both zero,

$$\begin{vmatrix} (\alpha v + n')^2 + \dfrac{1}{2}(v + n')^2, & \dfrac{3}{2}(v + n')^2 \\[3mm] \dfrac{3}{2}(v + n')^2, & (2v - \alpha v + n')^2 + \dfrac{1}{2}(v + n')^2 \end{vmatrix} = 0$$

or

(I.9) $$(\alpha - 1)^2 (\alpha + p)(\alpha - 2 - p) = 0,$$

where $p = n'/v$.

Corresponding to the root $\alpha = -p$ we have $B = -\overline{A}/3$, so that if $A = P + iQ$, P and Q being real disposable constants,

(I.10) $$w = (P + iQ)z^{-p} - \frac{1}{3}(P - iQ)z^{2+p},$$

and $\alpha = 2 + p$ leads to nothing new.

Corresponding to the double root $\alpha = 1$ we try

$$w = (C + inEt)z,$$

which leads to

$$\left[-\frac{3}{2}(v + n')^2 (C + \overline{C}) - 2v(v + n')E \right.$$

$$\left. -\frac{3}{2}(v + n')^2 in(E - \overline{E})t \right] z = 0$$

so that E must be real, and equal to $-3(1 + p)(C + \overline{C})/4$. So if $C = R + iS$, R and S being real disposable constants,

$$E = -\frac{3}{2}(1 + p)R,$$

and the corresponding part of the complementary function is

(I.11) $$w = \left(R + iS - \frac{3}{2}in(1 + p)Rt \right) z.$$

For a particular solution we try

$$w = A'z^{-1} + B'z^3 + (C' + intE')z,$$

which leads to

(I.12)
$$A' = \frac{3m^2 a(3p^2 + 14p + 19)}{16(p+3)(p-1)}$$
$$= b_{-1}, \text{ say, where } m = n'/\nu,$$

(I.13)
$$B' = \frac{-9m^2 a(p+1)^2}{16(p+3)(p-1)}$$
$$= b_3, \text{ say,}$$

(I.14)
$$E' = \frac{-m^2 a}{4(1+p)}$$
$$= c_1, \text{ say,}$$

and we may take $C' = 0$.
So the complete solution of the equation (I.8) is

(I.15)
$$w = (P + iQ)z^{-p} - \frac{1}{3}(P - iQ)z^{2+p}$$
$$+ \left\{ R + iS + \left[c_1 - \frac{3}{2}(1+p)R \right] i\nu t \right\} z + b_{-1}z^{-1} + b_3 z^3.$$

c. *The search for a periodic solution.* Suppose that at $t = 0$ we have

$$w = \beta_1 + i\beta_2 \text{ and } \dot{w} = \beta_3 + i\beta_4,$$

which corresponds to the general set of initial conditions

(I.16) $\xi = a + \beta_1, \quad \eta = \beta_2, \quad \dot{\xi} = \beta_3, \quad \dot{\eta} = a\nu + \beta_4.$

This corresponds to

$$\beta_1 = \frac{2}{3}P + R + b_{-1} + b_3,$$

$$\beta_2 = \frac{4}{3}Q + S,$$

$$\beta_3 = -\frac{2}{3}(1-p)\nu Q - \nu S,$$

and
$$\beta_4 = -\frac{2}{3}(1+2p)\nu P - \frac{1}{2}(1+3p)\nu R + \nu c_1 - \nu b_{-1} + 3\nu b_3.$$

Solving for the disposable constants,

$$P = -\frac{3(1+3p)}{2(1+p)}\,\beta_1' - \frac{3}{(1+p)\nu}\,\beta_4',$$

$$Q = \frac{3}{2(1+p)}\,\beta_2 + \frac{3}{2(1+p)\nu}\,\beta_3,$$

$$R = \frac{2(1+2p)}{1+p}\,\beta_1' + \frac{2}{(1+p)\nu}\,\beta_4',$$

and

$$S = -\frac{(1-p)}{1+p}\,\beta_2 - \frac{2}{(1+p)\nu}\,\beta_3,$$

where $\beta_1' = \beta_1 - b_{-1} - b_3$ and $\beta_4' = \beta_4 - \nu c_1 + \nu b_{-1} - 3\nu b_3$.

Suppose now that at $t = 2\pi/\nu + \tau$ we have

$$w = \beta_1 + \psi_1 + i(\beta_2 + \psi_2)$$

and

$$\dot{w} = \beta_3 + \psi_3 + i(\beta_4 + \psi_4).$$

There is a relation between the ψ_i, which arises from Jacobi's integral

(I.17) $$\frac{1}{2}\,(\dot{\xi}^2 + \dot{\eta}^2) - \frac{\mu}{r} + \frac{3}{2}\,n'^2\xi^2 = \text{constant}.$$

Substituting the expression (I.7) gives

$$\frac{1}{2}\,(a^2\nu^2 - 2a\nu\sin\nu t\delta\dot{\xi} + 2a\nu\cos\nu t\delta\dot{\eta})$$

$$-\frac{\mu}{a} + \frac{\mu}{a^2}\,(\cos\nu t\delta\xi + \sin\nu t\delta\eta)$$

$$+\frac{3}{2}\,n'^2(a^2 + 2a\cos\nu t\delta\xi)$$

$$+ O(\delta\xi^2, \delta\eta^2) = \text{constant},$$

and consideration of the values at $t = 0$ and $t = 2\pi/\nu + \tau$ shows that

$$a\nu\sin\nu\tau\psi_3 + a\nu\cos\nu\tau\psi_4 + (\nu + n')^2 a(\cos\nu\tau\psi_2 + \sin\nu\tau\psi_2)$$

$$+ 3n'^2 a\cos\nu\tau\psi_1 = 0.$$

Therefore it is sufficient, in order to establish a periodic orbit, to satisfy

(I.18) $\psi_1 = \psi_2 = \psi_3 = 0,$

since $\psi_4 = 0$ is then assured. We may then take β_2 and τ both to be zero, leaving β_1, β_3 and β_4 to be determined. The equations to be solved are then

$$\psi_1 = \frac{2}{3}(\cos 2\pi p - 1)P + \frac{2}{3}\sin 2\pi p \cdot Q = 0,$$

$$\psi_2 = -\frac{4}{3}\sin 2\pi p \cdot P + \frac{4}{3}(\cos 2\pi p - 1)Q + 2\pi\left[c_1 - \frac{3}{2}(1+p)R\right]$$

$$= 0,$$

and

$$\psi_3 = \frac{2}{3}\nu(1-p)\sin 2\pi p \cdot P - \frac{2}{3}\nu(1-p)(\cos 2\pi p - 1)Q$$

$$- 2\pi\nu\left[c_1 - \frac{3}{2}(1+p)R\right] = 0,$$

which lead to

$$P = Q = 0, \qquad \text{and} \qquad R = \frac{2c_1}{3(1+p)},$$

which in turn require that

$$\beta_3 = 0,\ \beta_1' = \frac{2c_1}{3(1+p)} \qquad \text{and} \qquad \beta_4' = -\frac{(1+3p)c_1}{3(1+p)}.$$

Then $S = 0$, and so the appropriate solution is (using $p = m$)

$$w = \frac{2c_1}{3(1+p)}z + b_{-1}z^{-1} + b_3 z^3$$

(I.19)

$$= -\frac{1}{6}am^2 z - \frac{19}{16}am^2 z^{-1} + \frac{3}{16}am^2 z^3 + O(m^3).$$

The Jacobian determinant, which must not vanish if the expression for the β_i appropriate to the periodic orbit are to be obtained as power series in m by the method indicated by Poincaré, is

$$\left(\frac{\partial(\psi_1, \psi_2, \psi_3)}{\partial(\beta_1, \beta_3, \beta_4)}\right)_{\tau=\beta_i=0} = \frac{6\pi}{n}(1 - \cos 2\pi p).$$

Correct to m^2, the expressions for the coordinates are, then,

$$\xi = a \left(1 - \frac{5}{4} m^2\right) \cos \nu t + \frac{3}{16} a m^2 \cos 3\nu t,$$

(I.20)

$$\eta = a \left(1 + \frac{9}{8} m^2\right) \sin \nu t + \frac{3}{16} a m^2 \sin 3\nu t.$$

This orbit crosses the axis of ξ at right angles, and is symmetric about it. It is, in fact, the intermediary orbit used by Hill in his Lunar Theory [1].

II. The equations of linear variation.

a. *General formulation.* Suppose a dynamical system is described by means of variables $x_1, x_2, \cdots, x_n$ which satisfy equations of the type

(II.1) $\qquad \dfrac{dx_i}{dt} = X_i(x_1, x_2, \cdots, x_n) \qquad (i = 1, 2, \cdots, n)$

which possess a periodic solution given by

(II.2) $\qquad\qquad x_i = \phi_i(t) \qquad (i = 1, 2, \cdots, n)$

which has period T. Then for motion in the vicinity of this solution

(II.3) $\qquad\qquad x_i = \phi_i(t) + \xi_i \qquad (i = 1, 2, \cdots, n)$

where the ξ_i are small. Substitution in (II.1) gives

(II.4) $\qquad \dfrac{d\xi_i}{dt} = \sum_{j=1}^{n} \dfrac{\partial X_i}{\partial \xi_j} \, (\phi_1(t), \phi_2(t), \cdots, \phi_n(t)) \, \xi_j + O(\xi^2)$

or, to first order in the ξ_i,

(II.5) $\qquad \dfrac{d\xi_i}{dt} = \sum_{j=1}^{n} A_{ij}(t) \, \xi_j \qquad (i = 1, 2, \cdots, n)$

where $A_{ij}(t) = (\partial X_i / \partial \xi_j)(\phi_1(t), \phi_2(t), \cdots, \phi_n(t))$. These are called the "equations of linear variation."

Now if $\xi_i^{(1)}(t), \xi_i^{(2)}(t), \cdots, \xi_i^{(n)}(t)$ are a linearly independent set of solutions of the equations (II.5), any solution may be written in the form

(II.6) $\qquad \xi_i(t) = \sum_{j=1}^{n} B^{(j)} \xi_i^{(j)}(t) \qquad (i = 1, 2, \cdots, n)$

where the $B^{(j)}$ are appropriate constants. Now the $\phi_i(t)$ have period T, and therefore so does $A_{ij}(t)$. So if $\xi_i = \xi_i^*(t)$ is a solution of (II.5), so also is $\xi_i = \xi_i^*(T+t)$. In particular so is $\xi_i = \xi_i^{(j)}(T+t)$ which may therefore be written, as in (II.6),

$$(II.7) \qquad \xi_i^{(j)}(T+t) = \sum_{k=1}^{n} B^{(j,k)}\xi_i^{(k)}(t) \qquad (i = 1, 2, \cdots, n)$$

for appropriate $B^{(j,k)}$.

We now seek a solution $\hat{\xi}_i(t)$ which has the property that

$$(II.8) \qquad \hat{\xi}_i(T+t) = \lambda\hat{\xi}_i(t) \qquad (i = 1, 2, \cdots, n), \text{ for all } t.$$

If $\hat{\xi}_i(t)$ is written in the form

$$\hat{\xi}_i(t) = \sum_{j=1}^{n} B^{(j)}\xi_i^{(j)}(t) \qquad (i = 1, 2, \cdots, n)$$

then this requires that

$$\sum_{j=1}^{n} B^{(j)} \cdot \xi_i^{(j)}(T+t) = \lambda\sum_{j=1}^{n} B^{(j)}\xi_i^{(j)}(t) \qquad (i = 1, 2, \cdots, n)$$

or, using (II.7), that

$$\sum_{j=1}^{n}\sum_{k=1}^{n} B^{(j)}B^{(j,k)}\xi_i^{(k)}(t) = \lambda\sum_{j=1}^{n} B^{(j)}\xi_i^{(j)}(t) \qquad (i = 1, 2, \cdots, n).$$

The $\xi_i^{(j)}(t)$ are a linearly independent set, and so we must have

$$\sum_{k=1}^{n} B^{(k)}B^{(k,j)} = \lambda B^{(j)} \qquad (j = 1, 2, \cdots, n).$$

Thus λ must be an eigenvalue of the matrix $B^{(k,j)}$, that is a solution of the equation

$$(II.9) \qquad \begin{vmatrix} B^{(1,1)} - \lambda, & B^{(1,2)} & , & \cdots, & B^{(1,n)} \\ B^{(2,1)} & , & B^{(2,2)} - \lambda, & \cdots, & B^{(2,n)} \\ \cdot & & \cdot & & \cdot \\ \cdot & & \cdot & & \cdot \\ \cdot & & \cdot & & \cdot \\ B^{(n,1)} & , & B^{(n,2)} & , & \cdots, & B^{(n,n)} - \lambda \end{vmatrix} = 0,$$

and $(B^{(1)}, B^{(2)}, \cdots, B^{(n)})$ is a corresponding eigenvector.

If λ is such an eigenvalue, then the value of α satisfying

$$(\text{II.10}) \qquad \lambda = \exp(\alpha T)$$

and whose imaginary part, $\mathscr{J}(\alpha)$, is in the range $-\pi/T \leq \mathscr{J}(\alpha) \leq \pi/T$, is called a "characteristic exponent" of the solution (II.2).

If the n solutions of the equation (II.9) are different, then the n corresponding eigenvalues are linearly independent, and the n corresponding solutions $\xi_i = \hat{\xi}_i(t)$ are linearly independent, and so form a complete set. So if the eigenvalues are $\lambda^{(1)}, \lambda^{(2)}, \cdots, \lambda^{(n)}$, the corresponding characteristic exponents are $\alpha^{(1)}, \alpha^{(2)}, \cdots, \alpha^{(n)}$, and the corresponding solutions are $\hat{\xi}_i^{(1)}(t), \hat{\xi}_i^{(2)}(t), \cdots, \hat{\xi}_i^{(n)}(t)$ respectively, then any solution may be written in the form

$$(\text{II.11}) \qquad \xi_i(t) = \sum_{k=1}^{n} C_k \hat{\xi}_i^{(k)}(t) \qquad (i = 1, 2, \cdots, n)$$

where the C_k are constants to be determined. Since $\hat{\xi}_i^{(k)}(t)$ satisfies (II.8) if the corresponding characteristic exponent is $\alpha^{(k)}$, we have, in view of (II.10), the result that the function

$$(\text{II.12}) \qquad \chi_i^{(k)}(t) = \exp(-\alpha^{(k)}t)\, \hat{\xi}_i^{(k)}(t)$$

for $i = 1, 2, \cdots, n$, is periodic with period T. Therefore the general solution (II.11) of the equations (II.5) may be written in the form

$$(\text{II.13}) \qquad \xi_i(t) = \sum_{k=1}^{n} C_k \exp(\alpha^{(k)}t)\chi_i^{(k)}(t) \qquad (i = 1, 2, \cdots, n)$$

where the $\chi_i^{(k)}(t)$ are periodic functions with the period T of the solution (II.2).

If two or more of the solutions of equation (II.9) are equal, then the form (II.13) does not usually give the general solution of the equations (II.5). Suppose the two of the solutions are equal, so that there are $n-1$ distinct eigenvalues, $\lambda^{(1)}, \lambda^{(2)}, \cdots, \lambda^{(n-1)}$, corresponding to which there are $n-1$ linearly independent solutions, $\hat{\xi}_i^{(1)}(t), \hat{\xi}_i^{(2)}(t), \cdots, \hat{\xi}_i^{(n-1)}(t)$, respectively, of (II.5). These do not form a complete set, so we can find a solution, say $\xi_i^*(t)$, which is not a linear combination of them. Now the set $\hat{\xi}_i^{(1)}(t), \hat{\xi}_i^{(2)}(t), \cdots, \hat{\xi}_i^{(n-1)}(t)$, $\xi_i^*(t)$ are linearly independent, n in number, and so form a complete set, and so the solution $\xi_i^*(T+t)$ may be written in terms of them as

$$(\text{II.14}) \qquad \xi_i^*(T+t) = \sum_{k=1}^{n-1} b_k \hat{\xi}_i^{(k)}(t) + b_n \xi_i^*(t).$$

Now consider the solution

(II.15) $$\hat{\xi}_i^{(n)}(t) = \sum_{k=1}^{n-1} c_k \hat{\xi}_i^{(k)}(t) + c_n \xi_i^*(t),$$

where the constants $c_1, c_2, \cdots, c_n$ are still to be determined. We have, since the solutions $\hat{\xi}_i^{(1)}(t), \cdots, \hat{\xi}_i^{(n-1)}(t)$ satisfy (II.8), and from (II.14),

$$\hat{\xi}_i^{(n)}(T+t) = \sum_{k=1}^{n-1} c_k \lambda^{(k)} \hat{\xi}_i^{(k)}(t) + c_n \left\{ \sum_{k=1}^{n-1} b_k \hat{\xi}_i^{(k)}(t) + b_n \xi_i^*(t) \right\}$$

$$= \sum_{k=1}^{n-1} (c_k \lambda^{(k)} + c_n b_k) \hat{\xi}_i^{(k)}(t) + b_n \left\{ \hat{\xi}_i^{(n)}(t) - \sum_{k=1}^{n-1} c_k \hat{\xi}_i^{(k)}(t) \right\},$$

(using (II.15) again), i.e.,

(II.16) $\hat{\xi}_i^{(n)}(T+t) = \sum_{k=1}^{n-1} \left\{ c_k(\lambda^{(k)} - b_n) + c_n b_k \right\} \hat{\xi}_i^{(k)}(t) + b_n \hat{\xi}_i^{(n)}(t).$

Suppose, if possible, that b_n is different from all of the $\lambda^{(k)}$. Then we may choose $c_k = c_n/(b_n - \lambda^{(k)})$ $(k = 1, 2, \cdots, n-1)$, and (II.16) becomes

$$\hat{\xi}_i^{(n)}(T+t) = b_n \hat{\xi}_i^{(n)}(t),$$

showing that $\hat{\xi}_i^{(n)}(t)$ satisfies (II.8), and $\lambda = b_n$ must be a solution of equation (II.9), contradicting the assumption that b_n is different from all of the roots of that equation. Therefore b_n must equal one of $\lambda^{(k)}$, and by appropriate enumeration of the latter we may take

$$b_n = \lambda^{(n-1)}.$$

Also we may take $c_k = c_n b_k/(\lambda^{(n-1)} - \lambda^{(k)})$, $k = 1, 2, \cdots, n-2$, and (II.16) becomes

(II.17) $$\hat{\xi}_i^{(n)}(T+t) = c_n b_{n-1} \hat{\xi}_i^{(n-1)}(t) + \lambda^{(n-1)} \hat{\xi}_i^{(n)}(t).$$

Now if $b_{n-1} = 0$, the functions

$$\chi_i^{(n)}(t) = \exp\left(-\alpha^{(n-1)}t\right) \hat{\xi}_i^{(n)}(t)$$

have period T, and the general solution has the form (II.13), as in the case where all the $\lambda^{(k)}$ are different, except that now $\alpha^{(n)} = \alpha^{(n-1)}$. If on the other hand $b_{n-1} \neq 0$, we put $c_n = 1/b_{n-1}$, and the function

$$\chi_i^{(n)}(t) = \exp\left(-\alpha^{(n-1)}t\right) \left\{ \hat{\xi}_i^{(n)}(t) - \frac{t}{T\lambda^{(n-1)}} \hat{\xi}_i^{(n-1)}(t) \right\}$$

is periodic with the period T. The general solution of the equations (II.5) now has the form[3]

$$\xi_i(t) = \sum_{k=1}^{n-1} C_k \exp(\alpha^{(k)} t) \chi_i^{(k)}(t)$$

(II.18)
$$+ C_n \exp(\alpha^{(n-1)} t) \left\{ \chi_i^{(n)}(t) + \frac{t}{T\lambda^{(n-1)}} \chi_i^{(n-1)}(t) \right\}$$

$$(i = 1, 2, \cdots, n).$$

In the case of a greater multiplicity of roots of (II.9), suppose the root $\lambda^{(s)}$ is repeated $p+1$ times, then we shall find that the corresponding contribution to the general solution is

$$\exp(\alpha^{(s)} t) \Bigg[C_s \chi_i^{(s)}(t) + C_{s+1} \chi_i^{(s+1)}(t) + \cdots + C_q \chi_i^{(q)}(t)$$

$$+ C_{q+1} \left\{ \chi_i^{(q+1)}(t) + \frac{t}{T\lambda^{(s)}} \chi_i^{(q)}(t) \right\}$$

$$+ C_{q+2} \left\{ \chi_i^{(q+2)}(t) + \frac{t}{T\lambda^{(s)}} \chi_i^{(q+1)}(t) \right.$$

(II.19)
$$\left. + \frac{t(t-T)}{2!(T\lambda^{(s)})^2} \chi_i^{(q)}(t) \right\} + \cdots$$

$$+ C_{s+p} \left\{ \chi_i^{(s+p)}(t) + \frac{t}{T\lambda^{(s)}} \chi_i^{(s+p+1)}(t) + \cdots \right.$$

$$\left. + \frac{t(t-T)(t-2T)\cdots(t-rT)}{(r+1)!(T\lambda^{(s)})^{r+1}} \chi_i^{(q)}(t) \right\} \Bigg]$$

where the functions $\chi_i^{(j)}(t)$ have period T, and q is some number which will lie in the range $s \leqq q \leqq s + p$, and $r = s + p - q - 1$.

b. *The case of constant coefficients.* If we are investigating motion in the vicinity of a configuration of equilibrium, then the quantities $\phi_i(t)$ of (II.2) are constants, and so the coefficients A_{ij} in the equations (II.5) are also constants.

Suppose that the eigenvalues of the matrix (A_{ij}), that is, the roots $x = x^{(1)}, x^{(2)}, \cdots, x^{(n)}$ of the equation

---

[3]The proof previously given for this result does not cover all cases. I am grateful to Dr. Barrar for casting doubt on it.

$$
(II.20) \quad
\begin{vmatrix}
A_{11} - x, & A_{12}, & \cdots, & A_{1n} \\
A_{21}, & A_{22} - x, & \cdots, & A_{2n} \\
\cdot & \cdot & & \cdot \\
\cdot & \cdot & & \cdot \\
\cdot & \cdot & & \cdot \\
A_{n1}, & A_{n2}, & \cdots, & A_{nn} - x
\end{vmatrix} = 0,
$$

are different. Then corresponding to each $x^{(j)}$ we can find an eigen-vector $(p_{1j}, p_{2j}, \cdots, p_{nj})$ such that

$$
(II.21) \quad \sum_{k=1}^{n} A_{ik} p_{kj} = x^{(j)} p_{ij} \qquad (i, j = 1, 2, \cdots, n).
$$

Then the n eigenvectors are linearly independent, and the matrix (p_{ij}) is consequently nonsingular. If $(\breve{p}_{ij})$ is its inverse matrix, define new variables $\eta_1, \eta_2, \cdots, \eta_n$ by

$$
(II.22) \quad \eta_i = \sum_{j=1}^{n} \breve{p}_{ij} \xi_j \qquad (i = 1, 2, \cdots, n).
$$

Then they satisfy the equations, using (II.5),

$$
\frac{d\eta_i}{dt} = \sum_{j=1}^{n} \breve{p}_{ij} \frac{d\xi_j}{dt}
$$

$$
= \sum_{j=1}^{n} \sum_{k=1}^{n} \breve{p}_{ij} A_{jk} \xi_k
$$

$$
= \sum_{j=1}^{n} \sum_{k=1}^{n} \sum_{l=1}^{n} \breve{p}_{ij} A_{jk} p_{kl} \eta_l,
$$

using the converse of (II.22). Then (II.21) shows that

$$
(II.23) \quad
\begin{aligned}
\frac{d\eta_i}{dt} &= \sum_{j=1}^{n} \sum_{l=1}^{n} \breve{p}_{ij} p_{jl} x^{(l)} \eta_l \\
&= x^{(i)} \eta_i \qquad (i = 1, 2, \cdots, n).
\end{aligned}
$$

This has the solution

$$
(II.24) \quad \eta_i(t) = \eta_i(0) \exp(x^{(i)} t) \qquad (i = 1, 2, \cdots, n),
$$

and therefore the solution of (II.5) is given in the case by

(II.25) $\xi_i(t) = \sum_{j=1}^{n} p_{ij} \eta_j(0) \exp{(x^{(j)}t)}$ $(i = 1, 2, \cdots, n)$.

The eigenvalues $x^{(j)}$ of (A_{ij}) are therefore to be identified with the characteristic exponents, and the periodic functions $\chi_i^{(j)}(t)$ are replaced by the constants p_{ij}.

c. *The equation satisfied by the characteristic exponents.* In Poincaré's notation [3, §§37-39, 11, 40, 42, 43 and 46-48], we have

(II.26)   and $\left.\begin{array}{l} \xi_i(0) = \beta_i \\ \xi_i(T) = \beta_i + \psi_i \end{array}\right\}$ $(i = 1, 2, \cdots, n)$,

and, since the ψ_i are zero when the β_i are, we may put

(II.27) $$\psi_i = \sum_{k=1}^{n} \left(\frac{\partial \psi_i}{\partial \beta_k}\right)_0 \beta_k + (\beta_i^2),$$

for a sufficiently small range of the β_i, where the suffix "0" indicates that the β_i have been put equal to zero. But if α is a characteristic exponent we know that the equations (II.5) have a solution of the form

(II.28) $\xi_i(t) = \epsilon \exp{(\alpha t)} \chi_i(t)$ $(i = 1, 2, \cdots, n)$,

where ϵ is a constant and the functions χ_i have period T. Making use of the relations (II.26),

$\beta_i = \epsilon \chi_i(0)$ and

(II.29) $\beta_i + \psi_i = \epsilon \exp{(\alpha T)} \chi_i(T)$

$= \epsilon \exp{(\alpha T)} \chi_i(0)$ $(i = 1, 2, \cdots, n)$.

Therefore the equation (II.27) may be written

$$\epsilon \{\exp{(\alpha T)} - 1\} \chi_i(0) = \sum_{k=1}^{n} \left(\frac{\partial \psi_i}{\partial \beta_k}\right)_0 \epsilon \chi_k(0) + O(\epsilon^2)$$

so that, dividing by ϵ and letting ϵ tend to zero,

(II.30) $\{\exp{(\alpha T)} - 1\} \chi_i(0) = \sum_{k=1}^{n} \left(\frac{\partial \psi_i}{\partial \beta_k}\right)_0 \chi_k(0)$ $(i = 1, 2, \cdots, n)$.

Since the $\chi_i(0)$ are not all zero, the quantity $x = \exp{(\alpha T)} - 1$ must be an eigenvalue of the matrix $(\partial \psi_i / \partial \beta_k)_0$, that is, a solution of the equation

$$\text{(II.31)} \quad \begin{vmatrix} \left(\dfrac{\partial\psi_1}{\partial\beta_1}\right)_0 - x, & \left(\dfrac{\partial\psi_1}{\partial\beta_2}\right)_0, & \cdots, & \left(\dfrac{\partial\psi_1}{\partial\beta_n}\right)_0 \\[2mm] \left(\dfrac{\partial\psi_2}{\partial\beta_1}\right)_0, & \left(\dfrac{\partial\psi_2}{\partial\beta_2}\right)_0 - x, \cdots, & & \left(\dfrac{\partial\psi_2}{\partial\beta_n}\right)_0 \\[2mm] \vdots & \vdots & & \vdots \\[2mm] \left(\dfrac{\partial\psi_n}{\partial\beta_1}\right)_0, & \left(\dfrac{\partial\psi_n}{\partial\beta_2}\right)_0, & \cdots, & \left(\dfrac{\partial\psi_n}{\partial\beta_n}\right)_0 - x \end{vmatrix} = 0$$

or

$$\det \left| \left(\frac{\partial\psi_i}{\partial\beta_j}\right)_0 - x\delta_{ij} \right| = 0,$$

where $\delta_{ij} = 0$ if $i \neq j$, and 1 if $i = j$. If the characteristic exponents are different they are all given by the n roots of this equation.

d. *Some properties of the characteristic exponents.* If the equations of motion (II.1) are real, then real values of the β_i lead to real values of the ψ_i, so that the $(\partial\psi_i/\partial\beta_k)_0$ are real, and if a root of equation (II.31) is complex, its complex conjugate is also a root. Therefore if a characteristic exponent is complex, its conjugate is also a characteristic exponent.

If the variables ξ_i are subjected to a linear transformation

$$\text{(II.32)} \qquad \eta_i = \sum_{j=1}^{n} T_{ij}(t)\,\xi_j \qquad (i = 1, 2, \cdots, n)$$

where the $T_{ij}(t)$ have period T, then if

$$\eta_i(0) = \beta_i', \quad \text{and} \quad \eta_i(T) = \beta_i' + \psi_i' \qquad (i = 1, 2, \cdots, n)$$

we have

$$\beta_i' = \sum_{j=1}^{n} T_{ij}(0)\,\beta_j$$

and

$$\beta_i' + \psi_i' = \sum_{j=1}^{n} T_{ij}(T)(\beta_i + \psi_i) \qquad (i = 1, 2, \cdots, n)$$

so that

(II.33) $\qquad \psi_i' = \sum_{j=1}^{n} T_{ij}(0)\, \psi_j \qquad (i = 1, 2, \cdots, n)$

and

(II.34) $\qquad \beta_i = \sum_{j=1}^{n} \check{T}_{ij}(0)\, \beta_j' \qquad (i = 1, 2, \cdots, n)$

where $(\check{T}_{ij})$ is the inverse matrix to (T_{ij}). The equation satisfied by the quantity $x' = \exp(\alpha'\, T) - 1$, where α' is a characteristic exponent derived from the equations for $d\eta_i/dt$, is

(II.35) $\qquad \det \left| \left(\dfrac{\partial \psi_i'}{\partial \beta_j'} \right)_0 - x'\delta_{ij} \right| = 0.$

But, using (II.33) and (II.34),

$$\left(\frac{\partial \psi_i'}{\partial \beta_j'} \right)_0 = \sum_{k=1}^{n} T_{ik}(0)\, \left(\frac{\partial \psi_k}{\partial \beta_j'} \right)_0$$

$$= \sum_{k=1}^{n} T_{ik}(0) \sum_{l=1}^{n} \left(\frac{\partial \psi_k}{\partial \beta_l} \right)_0 \check{T}_{lj}(0),$$

and since

$$\delta_{ij} = \sum_{k=1}^{n} T_{ik}(0)\, \delta_{kl}\, \check{T}_{lj}(0),$$

equation (II.35) may be written

$$\det | T_{ik}(0) | \cdot \det \left| \left(\frac{\partial \psi_k}{\partial \beta_l} \right)_0 - x'\delta_{kl} \right| \cdot \det | \check{T}_{lj}(0) | = 0,$$

or

(II.36) $\qquad \det \left| \left(\dfrac{\partial \psi_k}{\partial \beta_l} \right)_0 - x'\delta_{kl} \right| = 0,$

since

$$\det | \check{T}_{ij} | = 1 \div \det | T_{ij} |.$$

The equation (II.36) is the same as the equation (II.31) whose roots are the quantities x derived from the characteristic exponents arising from the ξ_i, which are therefore identical with those arising from the η_i. Therefore the characteristic exponents are unaltered when a linear transformation of the type (II.32) is applied to the variables ξ_i.

If the variables x_i of the equations (II.1) are subjected to a general transformation

(II.37) $y_i = f_i(x_1, x_2, \cdots, x_n)$ $(i = 1, 2, \cdots, n)$

then the periodic solution (II.2) is described by the values $y_i = f_i\{\phi_1(t), \phi_2(t), \cdots, \phi_n(t)\}$ $(i = 1, 2, \cdots, n)$ of the y_i. The equations of linear variation for the y_i are obtained by putting

$$y_i = f_i\{\phi_1(t), \cdots, \phi_n(t)\} + \eta_i \qquad (i = 1, 2, \cdots, n)$$

corresponding, to first order in the η_i, to the motion described by (II.3), namely

$$x_i = \phi_i(t) + \xi_i \qquad (i = 1, 2, \cdots, n),$$

so that

(II.38) $\displaystyle \eta_i = \sum_{j=1}^{n} \frac{\partial f_i}{\partial x_j} \{\phi_1(t), \cdots, \phi_n(t)\} \xi_j.$

The coefficients in this last equation have the period T of the periodic solution (II.2), and so this transformation is of the type (II.32), and leaves the characteristic exponents unaltered. Therefore the characteristic exponents associated with a given periodic solution of a dynamical problem are independent of the variables used.

Suppose the equations (II.1) possess an integral of the form

(II.39) $F(x_1, x_2, \cdots, x_n) = C.$

Then in the motion (II.3) we must have

$$F(\phi_1 + \beta_1, \cdots, \phi_n + \beta_n) = F(\phi_1 + \beta_1 + \psi_1, \cdots, \phi_n + \beta_n + \psi_n),$$

and so, differentiating with respect to β_j,

$$\sum_{i=1}^{n} \frac{\partial F}{\partial x_i} \frac{\partial \psi_i}{\partial \beta_j} = 0 \qquad (j = 1, 2, \cdots, n).$$

Thus there is a linear relation between the columns of the matrix unless the solution (II.2) corresponds to a stationary value of F with respect to all the x_i, and so equation (II.31) has a zero root, and there is a zero characteristic exponent. It can be shown that there is a zero characteristic exponent corresponding to each independent integral of the type (II.39) [3, §65]. This result is true even if the functions X_i of the equations (II.1) depend explicitly on the time (with respect to which they must then of course have the period

T, if the solution (II.2) is to be possible).

If the functions X_i do not depend explicitly on the time, then a solution near to (II.2) is

$$x_i = \phi_i(t + \delta t) \qquad (i = 1, 2, \cdots, n),$$

where δt is an increment in the time. This corresponds to the solution

$$(II.40) \qquad \xi_i = \frac{d\phi_i}{dt} \delta t \qquad (i = 1, 2, \cdots, n)$$

of the equations of linear variation. This solution has period T, and so corresponds to a zero characteristic exponent, which must exist even in the absence of integrals of the type (II.39), and may be shown to be additional to the zero exponents arising from such integrals [**3**, §§66 and 67].

Again, if the solution (II.2) is a member of a one parameter family of such solutions, all of the same period,

$$x_i = \phi_i(\sigma, t) \qquad (i = 1, 2, \cdots, n)$$

then another solution is

$$x_i = \phi_i(\sigma + \delta\sigma, t) \qquad (i = 1, 2, \cdots, n)$$

corresponding to

$$(II.41) \qquad \xi_i = \frac{\partial \phi_i}{\partial \sigma} \delta\sigma \qquad (i = 1, 2, \cdots, n),$$

which is independent of (II.40), and therefore corresponds to another zero characteristic exponent.

e. *The case of a canonical system.* Consider now a dynamical system whose equations of motion may be put in the canonical form

$$(II.42) \qquad \frac{dx_i}{dt} = \frac{\partial H}{\partial y_i}, \qquad \frac{dy_i}{dt} = -\frac{\partial H}{\partial x_i} \qquad (i = 1, 2, \cdots, n)$$

where H is a function of the variables $x_1, y_1; x_2, y_2; \cdots; x_n, y_n$, of the problem. The variables may be arranged in conjugate pairs. Let

$$(II.43) \qquad x_i = \phi_i(t), \qquad y_i = \phi_{n+i}(t) \qquad (i = 1, 2, \cdots, n)$$

be a periodic solution, the functions $\phi_i(t)$ having period T. Then the equations of linear variation are obtained by putting

(II.44) $x_i = \phi_i(t) + \xi_i, \qquad y_i = \phi_{n+1}(t) + \eta_i \qquad (i = 1, 2, \cdots, n)$

and retaining only the first order in ξ_i and η_i, which gives

$$
\text{(II.45)} \quad
\begin{aligned}
\frac{d\xi_i}{dt} &= \sum_{j=1}^{n} \left\{ \frac{\partial^2 H}{\partial x_j \partial y_i} \xi_j + \frac{\partial^2 H}{\partial y_j \partial y_i} \eta_j \right\} \\
\frac{d\eta_i}{dt} &= \sum_{j=1}^{n} \left\{ \frac{\partial^2 H}{\partial x_j \partial x_i} \xi_j + \frac{\partial^2 H}{\partial y_j \partial x_i} \eta_j \right\}
\end{aligned}
\qquad (i = 1, 2, \cdots, n)
$$

the second derivatives of H being evaluated by use of the expressions (II.43). Then if (ξ_i, η_i) and (ξ_i', η_i') are two solutions of these equations we may easily verify that

$$
\sum_{i=1}^{n} (\xi_i \eta_i' - \xi_i' \eta_i)
$$

is constant.

If α and α' are two characteristic exponents of the solution then corresponding to them there exist solutions of the equations given by

$$
\begin{aligned}
\xi_i &= \exp(\alpha t)\, \chi_i(t) \\
\eta_i &= \exp(\alpha t)\, \chi_{n+i}(t)
\end{aligned}
\qquad (i = 1, 2, \cdots, n)
$$

and

$$
\begin{aligned}
\xi_i' &= \exp(\alpha' t)\, \chi_i'(t) \\
\eta_i' &= \exp(\alpha' t)\, \chi_{n+i}'(t)
\end{aligned}
\qquad (i = 1, 2, \cdots, n)
$$

where the functions $\chi_i(t)$ and $\chi_i'(t)$ either have period T, or are finite polynomials in t with coefficients which have period T, as we have seen in §IIa. Then the quantity

$$
\exp(\alpha + \alpha') \sum_{i=1}^{n} \left\{ \chi_i(t)\chi_{n+i}'(t) - \chi_i'(t)\chi_{n+i}(t) \right\},
$$

is constant, and so either $\alpha = -\alpha'$, or the constant is zero. That $(\alpha + \alpha')$ could be equal to an integral nonzero multiple of $2\pi i/T$ is excluded since the α are chosen to have imaginary parts in the range $-\pi/T$ to $+\pi/T$, and if $\alpha = \alpha' = \pi i/T$, then one of them may be taken equal to $-\pi i/T$ without altering the results of §IIa, and then we have $\alpha = -\alpha'$.

Suppose now that q of the characteristic exponents have the same value α. Then there are q linearly independent solutions corresponding to them, say

$$
\xi_i^{(r)} = \exp(\alpha t)\, \chi_i^{(r)}(t), \qquad \eta_i^{(r)} = \exp(\alpha t)\, \chi_{n+i}^{(r)}(t),
$$

where r takes the values $1, 2, \cdots, q$. Consider now those solutions (ξ_i, η_i) for which

$$(\text{II.46}) \qquad \sum_{i=1}^{n} (\xi_i \eta_i^{(r)} - \xi_i^{(r)} \eta_i) = 0,$$

for $r = 1, 2, \cdots, q - 1$ and q. There are q such relations, and so we may use them to reduce the system of $2n$ equations (II.45) by elimination to a system of degree $2n - q$, which therefore has $2n - q$ linearly independent solutions. But the system (II.45) has $2n$ independent solutions altogether, and so therefore q of these have

$$\sum_{i=1}^{n} (\xi_i \eta_i^{(r)} - \xi_i^{(r)} \eta_i) \neq 0$$

for some r, and therefore there are q independent solutions corresponding to a characteristic exponent $\alpha' = -\alpha$, so that q of the exponents have that value. Therefore, in a canonical system, to each characteristic exponent there corresponds an exponent equal in magnitude but opposite in sign.

f. *The stability of an equilibrium configuration or periodic motion.* A formal definition of the stability of a configuration was given by Liapounov [4], who called a configuration "stable" if, given any open neighborhood of phase space containing the configuration, then we can find another open neighborhood such that all motions originating within the second neighborhood always remain within the first one subsequently. Thus we may call the solution (II.2), namely

$$x_i = \phi_i(t) \quad (i = 1, 2, \cdots, n)$$

of the equations (II.1) stable if, given any positive number ϵ, we can find another positive number δ (in general depending on ϵ) such that, for any motion $x_i = x_i(t)$ satisfying

$$\sum_{i=1}^{n} |x_i(t_0) - \phi_i(t_0)| < \delta$$

for some time t_0, we have

$$\sum_{i=1}^{n} |x_i(t) - \phi_i(t)| < \epsilon$$

at all times t after t_0.

If the equations of linear variation described the equations of motion of the displaced motions (II.3) exactly, then their solution (II.13), (II.18) or (II.19) shows that the condition for stability is satisfied if all the characteristic exponents have negative real parts, or if they all have negative or zero real parts, those with zero real parts being distinct. If one at least has negative real part, and the others have zero real part, the displaced motion will in general tend asymptotically towards the solution (II.2) as the time tends to infinity. In this case the solution (II.2) is said to be "asymptotically stable". In the case of dynamical systems whose equations may be written in the canonical form (II.42), which is the case if the system is holonomic and the forces are conservative, which therefore applies in the case of the motion of celestial bodies under gravitational forces, we have seen that the characteristic exponents occur in pairs with opposite sign. Thus in this case, for stability, they must all be pure imaginary or zero, and different. Then the general solution of the equations of linear variation is a sum of sines and cosines of the time, and the definition of stability is satisfied.

In most cases, or course, the exact equations of motion of the displaced motion will have terms of higher degree in the ξ_i than the first, and the effect of these terms must be considered. Liapounov [4] showed that they cannot alter the fact that a solution which is shown by the equations of linear variation to be asymptotically stable is in fact asymptotically stable. Where all the exponents have zero real parts, and are distinct, however, these higher terms may prevent the foregoing formal definition of stability from being satisfied. Birkhoff [5, Chapter IV] showed that, in this case, nevertheless, if s is any integer greater than unity, there exists a finite number K_s such that, given any positive number ϵ (not larger than some definite quantity), then in all motions satisfying

$$\sum_{i=1}^{n} |x_i(t_0) - \phi_i(t_0)| < \epsilon$$

we have

$$\sum_{i=1}^{n} |x_i(t) - \phi_i(t)| < 2\epsilon$$

for all times satisfying $|t - t_0| < (1/K_s\epsilon^s)$, thus placing definite bounds on the rate at which the displaced motion departs from the solution (II.2).

In the case of periodic solutions, the above definition of stability is rarely satisfied even in the equations of linear variation; for example, in a one-parameter family of periodic solutions, the period usually varies with the parameter, and so in a disturbed motion whose initial conditions place it on a neighboring periodic solution to the original one, the point in phase space representing the motion is carried steadily further away from the corresponding point in the original motion, because of the progressive increase of the difference in phase, although the paths may always lie close together. The longitude of a planet or satellite may depart linearly from the corresponding value in the original motion, because of a small change in the mean motion arising from a small constant change in the major axis, without this constituting what we would wish to call an instability in the original motion. This shows itself in the term linear in time arising in the solution of the equations of linear variation when two of the characteristic exponents have equal values. We have seen that a periodic solution will usually have a number of zero characteristic exponents, which will usually give rise to such terms. We are therefore led to define the solution (II.2) as "orbitally stable" if any displaced motion, originating within an appropriate distance in phase space of some point in the original motion, has no point which is further than assigned distance from all points of the original path. Each case must be examined on its merits, to discover whether the powers of time arising in the solution of the equations of linear variation are allowable in this sense.

References

1. G. W. Hill, Amer. J. Math. 1 (1878), 5-26; 129-147; 245-260.

2. W. deSitter, Ann. Leiden Observatory 12(1918); part 1.

3. H. Poincaré, *Méthodes nouvelles de la mécanique céleste*, Gauthier-Villars, Paris, 1892.

4. A.M. Liapounov, *General problem of stability of motion*, Princeton Univ. Press, Princeton, N. J., 1947 (originally published in Russian, Kharkov, 1892).

5. G.D. Birkhoff, *Dynamical systems*, Amer. Math. Soc. Colloq. Publ. No. 9, Amer. Math. Soc., Providence, R.I., 1927; pp. 97-102.

UNIVERSITY OF LIVERPOOL

Paul B. Richards

Lectures
on Regularization

I. **Existence and uniqueness theorems of ordinary differential equations.** a. *Single Equation.* Consider the first order differential equation

$$(I.1) \qquad \frac{dx}{dt} = f(t, x)$$

where f is a real-valued function of the real variables t, x. Let (t_0, x_0) be a particular pair of values assigned to the variables (t, x).

PICARD-LINDELÖF THEOREM. HYPOTHESIS: 1. $f(t, x)$ *continuous in*

$$R: |t - t_0| \leqq a, |x - x_0| \leqq b;$$

2. $|f(t, x)| < A$ *(constant) in* R;

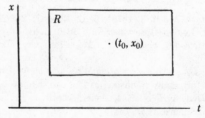

FIGURE 1. Domain for Solving (I.1)

100

3. $|f(t, x_1) - f(t, x_2)| \leq M|x_2 - x_1|$ *in* R, $M = constant$.

CONCLUSION: *There exists a unique function* $\phi(t)$, *defined for all values of* t *in* S: $|t - t_0| < h = \min(a, b/A)$, *such that* $x = \phi(t)$ *satisfies* (I.1) *and the initial condition,* $\phi(t_0) = x_0$.

Note. Condition 3 of the Hypothesis is called a Lipschitz condition.

This is a fundamental existence and uniqueness theorem in the theory of ordinary differential equations. Proofs of this theorem and others are found in standard texts on differential equations, such as [1], [2], [3], [4], [5], [6], and [7].

One method of proof, known as the Method of Successive Approximations, follows the Picard method of solution of the integral equation

$$x(t) = x_0 + \int_{t_0}^{t} f\{\tau, x(\tau)\} d\tau,$$

whose solution satisfies (I.1) and reduces to x_0 when $t = t_0$. Picard proved that in the interval S the sequence of functions

$$x^{(1)}(t) = x_0 + \int_{t_0}^{t} f\{\tau, x_0\} d\tau,$$

(I.2) $$x^{(2)}(t) = x_0 + \int_{t_0}^{t} f\{\tau, x^{(1)}(\tau)\} d\tau,$$

....

$$x^{(k)}(t) = x_0 + \int_{t_0}^{t} f\{\tau, x^{(k-1)}(\tau)\} d\tau,$$

converges to the desired solution.

The existence of the constant A in Condition 2 of the Hypothesis follows from Condition 1. Actually, however, the continuity of $f(t, x)$ is not necessary for the existence of a solution. Ince points out (see [1, p. 66]) that Mie has shown that solutions exist when $f(t, x)$ is continuous in x and Riemann-integrable in t. The Lipschitz condition is required to prove uniqueness. The significant point of this theorem is the narrow interval, S, of convergence that it guarantees. The solution may converge throughout a larger interval, but no general method of determining the exact boundaries of the interval of convergence has yet been discovered (see [1]). We say therefore that solutions of (I.1) hold in the "small," in general.

CARL A. RUDISILL LIBRARY
LENOIR RHYNE COLLEGE

On the other hand, if (I.1) is a linear differential equation, the interval, S, of convergence of the solution is the interval of continuity of the coefficients. Specifically, if $f(t, x) = ax + b$, where a, b are continuous functions of t in the interval $I = [t_1, t_2]$, then the solution exists and is unique in I (see [1], [6]). Thus in the case of linear differential equations we say the solutions hold in the "large."

b. *System of First-Order Equations.* Consider now the system of first-order equations

$$\frac{dx_1}{dt} = f_1(t, x_1, x_2, \cdots, x_n),$$

(I.3)   ....

$$\frac{dx_n}{dt} = f_n(t, x_1, x_2, \cdots, x_n).$$

The extension of the Picard-Lindelöf Theorem to this system is stated as follows:

HYPOTHESIS: 1. *Each f_i, $i = 1, 2, \cdots, n$, is continuous in its arguments in R:* $|t - t_0| \leqq a$, $|x_1 - x_1^0| \leqq b_1, \cdots, |x_n - x_n^0| \leqq b_n$.

2. $|f_i(t, x_1, \cdots, x_n)| < A$ *in* R, $i = 1, 2, \cdots, n$.

3. $|f_i(t, X_1, \cdots, X_n) - f_i(t, x_1, \cdots, x_n)|$
   $< K_1|X_1 - x_1| + \cdots + K_n|X_n - x_x|$ *in* $R, i = 1, 2, \cdots, n$.

CONCLUSION: *There exists a unique set of functions $\phi_i(t)$ defined for all values of t in S:* $|t - t_0| < h = \min(a, b_1/A, b_2/A, \cdots, b_n/A)$, *such that $x_i = \phi_i(t)$ satisfies* (I.3) *and the initial conditions,* $\phi_i(t_0) = x_i^0$.

The solutions are obtained by the recursion formula

(I.4) $$x_i^{(k)}(t) = x_i^{(0)} + \int_{t_0}^{t} f_i\{\tau, x_1^{(k-1)}(\tau), \cdots, x_n^{(k-1)}(\tau)\} \, d\tau,$$
$$i = 1, 2, \cdots, n,$$

in extension of (I.2).

Again, if (I.3) is a system of linear differential equations, i.e., if $f_i = a_{i1}x_1 + \cdots + a_{in}x_n + b_i$, $i = 1, \cdots, n$, where the coefficients $a_{i1}, \cdots, a_{in}$, and b_i are continuous functions of t in the interval $I = [t_1, t_2]$, then the set of solutions exists and is unique in I, and we say the solutions hold in the "large."

c. *Equations of Order Greater than One.* Finally, since the differential equation of order p,

CARL A. RUDISILL LIBRARY
LENOIR RHYNE COLLEGE

(I.5)
$$\frac{d^p x}{dt^p} = f\left(t, x, \frac{dx}{dt}, \cdots, \frac{d^{p-1} x}{dt^{p-1}}\right)$$

is equivalent to the set of p equations of the first order

$$\frac{dy_1}{dt} = y_2,$$

$$\frac{dy_2}{dt} = y_3,$$

(I.6) . . . .

$$\frac{dy_{p-1}}{dt} = y_p,$$

$$\frac{dy_p}{dt} = f(t, y_1, y_2, \cdots, y_p)$$

where we define

$$y_1 = x, y_2 = \frac{dx}{dt}, \cdots, y_p = \frac{d^{p-1} x}{dt^{p-1}},$$

it follows that if f is continuous and satisfies a Lipschitz condition in a domain R of dimension p, the equation (I.5) admits in a domain S a unique solution which, together with its first $p - 1$ derivatives, will assume an arbitrary set of initial conditions for the initial value $t = t_0$.

In the case of a linear differential equation of order $p > 1$,

(I.7)
$$a_0(t) \frac{d^p x}{dt^p} + a_1, (t) \frac{d^{p-1} x}{dt^{p-1}} + \cdots + a_{p-1}(t) \frac{dx}{dt}$$
$$+ a_p(t) x = b(t),$$

the equivalent set of p equations of the first order is

$$\frac{dy_1}{dt} = y_2, \cdots, \frac{dy_{p-1}}{dt} = y_p,$$

. . . .

$$\frac{dy_p}{dt} = \frac{b(t)}{a_0(t)} - \frac{a_p(t)}{a_0(t)} y_1 - \frac{a_{p-1}(t)}{a_0(t)} y_2 - \cdots - \frac{a_1(t)}{a_0(t)} y_p.$$

It follows that if $a_0(t)$, $a_1(t), \cdots, a_p(t)$ and $b(t)$ are continuous functions of t in $I = [t_1, t_2]$, and $a_0(t) \neq 0$ in I, then equation (I.7) admits of a unique solution in I which, together with its first $p - 1$ derivatives, assumes the assigned initial conditions for the initial value t_0 in I (see [1], [6]).

d. *Analytic Differential Equations.* A large class of physical problems are governed by equations of the form (I.1) where $f(t, x)$ is analytic in its arguments t and x in a region R: $|t - t_0| \leq a$, $|x - x_0| \leq b$, including the initial values of the variables. By this is meant $f(t, x)$ has continuous partial derivatives of all orders and is uniquely expansible in a power series in $(t - t_0)$ and $(x - x_0)$ which converges for all values of x and t in R. The power series is the Taylor series

$$f(t, x) = f(t_0, x_0) + \left(\frac{\partial f}{\partial x}\right)_{t_0, x_0} (x - x_0) + \left(\frac{\partial f}{\partial t}\right)_{t_0, x_0} (t - t_0)$$

$$+ \frac{1}{2!} \left\{ \left(\frac{\partial^2 f}{\partial x^2}\right)_{t_0, x_0} (x - x_0)^2 + 2 \left(\frac{\partial^2 f}{\partial x \partial t}\right)_{t_0, x_0} (x - x_0)(t - t_0)\right.$$

$$\left. + \left(\frac{\partial^2 f}{\partial t^2}\right)_{t_0, x_0} (t - t_0)^2\right\} + \cdots.$$

The Cauchy Existence Theorem applies to these analytic differential equations. It states (see [5]):

If $f(t, x)$ is analytic in its arguments in R: $|t - t_0| \leq a$, $|x - x_0| \leq b$, including the initial values of the variables, then the solution $x = \phi(t)$ of (I.1) is analytic in t, i.e., can be represented by a convergent power series in $(t - t_0)$, in a sufficiently small interval S about t_0.

The coefficients $c_1, c_2, \cdots, c_n, \cdots$ of the series solution

$$x(t) = x_0 + c_1(t - t_0) + c_2(t - t_0)^2 + \cdots$$

are determined by replacing x on both sides of (I.1) by this series, and then equating coefficients of like powers of $(t - t_0)$.

Since Cauchy, others[1] have extended the lower limits for S within which the solutions certainly converge, and Moulton (see [5, p. 27]) shows how one such S is automatically enlarged if f does not contain t explicitly (the differential equation is then said to be autonomous). In many cases the interval of convergence of the solution is larger than that predicted from the general theory, but there is no general

---

[1]See [5, p. 38] for specific references.

method of determining the exact interval of convergence of solutions of analytic nonlinear differential equations (see [5]), hence solutions of such equations hold in the "small." For example, consider the analytic nonlinear problem

$$\frac{dx}{dt} = 1 + x^2,$$

$$x(0) = 0.$$

Here the maximum interval of convergence predicted by the existence theorem is $|t| < 1/2$. On the other hand, direct integration is possible, and gives the solution $x = \tan t$, which has the interval of convergence $\pi/2$, approximately three times the predicted interval. The danger of assuming a large interval of convergence in the general case, where the exact solution is not known, is obvious.

As might be anticipated, the interval of convergence of solutions of analytic linear differential equations is the interval of convergence of the analytic coefficients (see [5]), and thus the solutions hold in the "large."

The Cauchy Theorem is easily extended to systems of analytic differential equations of order one and higher.

e. *Singularities and Regularization.* A singular point of a differential equation (I.1) is defined as a point (t, x) at which one or more of the conditions of the hypothesis of the existence theorem are not satisfied (see [1], [6]). Such points can be found by inspection of $f(t, x)$.

Thus, at a singular point of the differential equation, the existence theorem does not apply, and other methods must be sought for establishing the existence and nature of the solution. One such method is called regularization.

Regularization is essentially a substitution of an analytic (regular) problem for the singular one. It is applicable to equations with removable singularities and is accomplished by a transformation of variables which effect the removal of the singularity. Thus the solution of the regularized equation is the solution of the original equation.

An elementary example will illustrate regularization. Consider the problem

$$\frac{dx}{dt} = \frac{1}{x}, \qquad x \geqq 0, \qquad x(0) = 0.$$

This autonomous equation is singular at $x = 0$, the initial point, so the existence theorem does not apply in the neighborhood of this point $(t = 0, \ x = 0)$. Elsewhere the equation is analytic.

Introduce a new independent variable τ through the transformation $dt/d\tau = x$. Since $x \geq 0$, t is a monotone increasing function of τ, and $dt/d\tau = 0$ if and only if $x = 0$ $(t = 0)$. Hence, there is a one-to-one correspondence between t and τ for all $t \geq 0$, $\tau \geq \tau_0$, where $\tau = \tau_0$ corresponds to $t = 0$. The modified problem is

$$\frac{dx}{d\tau} = 1, \qquad x \geq 0,$$

$$x(\tau_0) = 0,$$

since

$$\frac{dx}{d\tau} = \frac{dx}{dt}\frac{dt}{d\tau}.$$

The solution, analytic for all finite τ, is

$$x = \tau - \tau_0.$$

To express x as a function of the original variable t, first obtain τ as a function of t from the equation $dt/d\tau = x$.
Integrating,

$$t = \int_{\tau_0}^{\tau} x\,d\tau = \int_{\tau_0}^{\tau} (\tau - \tau_0)\,d\tau = (\tau - \tau_0)^2/2,$$

from which

$$\tau - \tau_0 = (2t)^{1/2}, \qquad \tau \geq \tau_0.$$

Then the solution to the singular problem is

$$x = (2t)^{1/2}, \qquad t \geq 0,$$

analytic for $t > 0$.

In the next section we will discuss regularization of the differential equations of celestial mechanics.

II. **Regularization of the relative n-body equations of celestial mechanics—a formulation of the problem.** a. *Introduction.* In a gravitational force field containing n-bodies of mass m_i, $i = 1, 2, \cdots, n$, the equations of motion, relative to a nonrotating coordinate system whose origin is mass m_1, are given by the system of $n - 1$ vector equations (see [8]):

$$(II.1) \qquad \frac{d^2 \bar{\rho}_i}{dt^2} = - \frac{k^2 M_i}{\rho_i^3} \bar{\rho}_i + k^2 \sum_{j=2; j \neq i}^{n} m_j \left[\frac{\bar{\rho}_{ij}}{\rho_{ij}^3} - \frac{\bar{\rho}_j}{\rho_j^3} \right], \quad i = 2, 3, \cdots, n$$

where

$M_i = m_1 + m_i,$

k^2 is the gravitational constant,

$\bar{\rho}_{ij}$ is the vector from mass m_i to mass m_j,

$\bar{\rho}_i$ is the vector from the origin (mass m_1) to mass m_i,

$\bar{\rho}_j$ is the vector from the origin to mass m_j, and the real variable t denotes the time.

Except in the case $n = 2$, there are no known solutions of these equations which give qualitative information of the motion for all values of the time for arbitrary initial conditions.

The Cauchy Theorem establishes the existence of series solutions of equations (II.1) in a neighborhood of every initial time t_0 except possibly at collisions of two or more masses, which represent singularities of the differential equations (see [9]).

This suggests one approach to the problem—write the series solution at a point a short distance from a singularity, and by means of the process of analytic continuation derive successive sets of other power series which converge for values of the time successively closer to the singularity, but never reaching it, of course. By its construction, this approach precludes any survey of the general character of the motion near the collision.

Another theoretical approach, whose literature is much less abundant, is that of regularization of the equations of motion, i.e., removal of singularities. This would establish the existence of solutions at the singularities and might make it possible to determine the character of the motion near collisions.

Several investigators have succeeded in regularizing special cases of the n-body problem. G. D. Birkhoff and Levi-Civita have used conformal mapping techniques to remove singularities of the restricted 3-body equations. Levi-Civita removed one singularity by a clever choice of mapping, and Birkhoff succeeded in completely regularizing the restricted 3-body equations by another mapping (see [10]). Birkhoff used the new formulation to describe what he called the "states of motion," i.e., the position and velocity vectors of a particle at an instant.

Sundman in [11] regularized the general 3-body equations, except in the case of triple collision, by a change of the independent variable involving explicitly the position coordinates of the bodies. Sundman proved the existence of series solutions convergent for all values of time. Unfortunately, however, this is not of great practical value in studying the overall motion, because the particular series converges slowly for large time intervals (see [9], [12]).

b. *Transformed Equations.* Restricting motion to the invariable plane of the solar system, with mass m_1 the sun, Equations (II.1) have the same form, but can now be written in complex notation as

(II.2)

$$\frac{d^2 z_i}{dt^2} = -\frac{k^2 M_i}{|z_i|^3} z_i$$

$$+ k^2 \sum_{j=2; j \neq i}^{n} m_j \left[\frac{(z_j - z_i)}{|z_j - z_i|^3} - \frac{z_j}{|z_j|^3} \right], \qquad i = 2, 3, \cdots, n,$$

where

$$z = x(t) + iy(t).$$

Now consider the conformal map f of the z-plane (invariable plane) into the w-plane, and denote by g the inverse of f. Further, denote by the real variable τ the trajectory time parameter in the w-plane. Under this simultaneous conformal map and time transformation the equations (II.2) become

(II.3)

$$\frac{d^2 w_i}{d\tau^2} + \frac{g''(w_i)}{g'(w_i)} \left(\frac{dw_i}{d\tau} \right)^2 - \frac{d^2 t/d\tau^2}{dt/d\tau} \frac{dw_i}{d\tau}$$

$$= \frac{\left(\frac{dt}{d\tau} \right)^2}{g'(w_i)} \left\{ - \frac{k^2 M_i}{|g(w_i)|^3} g(w_i) \right.$$

$$\left. + k^2 \sum_{j=2; j \neq i}^{n} m_j \left[\frac{g(w_j) - g(w_i)}{|g(w_j) - g(w_i)|^3} - \frac{g(w_j)}{|g(w_j)|^3} \right] \right\},$$

$$i = 2, 3, \cdots, n,$$

where

$$w = u(\tau) + iv(\tau), \quad \text{and} \quad g'(w_i) = \frac{dg}{dw_i}.$$

The conformal property of the mapping specifies that

$$\left|\frac{dt}{d\tau}\right| = \left|g'(w_i)\right| \left|\frac{dw_i}{d\tau}\right| \Big/ \left|\frac{dz_i}{dt}\right|, \qquad i = 2, 3, \cdots, n,$$

where $|dw_i/d\tau|/|dz_i/dt|$ is the ratio of the speeds at corresponding points of trajectories in the w-plane and z-plane.

Of the four nonlinear terms in equations (II.3), the two that are carried over from equations (II.2) and exhibit the singularities of those differential equations are governed by the map g and by $dt/d\tau$. The nonlinear term

$$\frac{g''(w_i)}{g'(w_i)}\left(\frac{dw_i}{d\tau}\right)^2$$

is governed by the map g and is seen to vanish if and only if the transformation consists of a rotation, magnification and/or translation. It is unlikely, therefore, that (II.3) can be linearized as well as regularized. The presence in equations (II.3) of this new nonlinear term does not create additional singularities, since $g' \neq 0$, and hence does not affect essentially the existence problem.

In particular, under the identity transformation, equations (II.3) become

$$\frac{d^2 z_i}{d\tau^2} - \frac{d^2 t/d\tau^2}{dt/d\tau} \frac{dz_i}{d\tau}$$

$$(\text{II.4}) \qquad = \left(\frac{dt}{d\tau}\right)^2 \left\{ -\frac{k^2 M_i}{|z_i|^3} z_i + k^2 \sum_{j=2; j \neq i}^{n} m_j \left[\frac{z_j - z_i}{|z_j - z_i|^3} - \frac{z_j}{|z_j|^3}\right]\right\},$$

$$i = 2, 3, \cdots, n;$$

the equations of motion in the z-plane after a transformation in the independent variable.

With the selection of regularizing mappings, solutions of equations (II.3) may be written in the neighborhood of images of collisions. These solutions will immediately transform back to the physical plane via the selected mappings, without the necessity of solving equations (II.2).

Restriction of the dependent variable transformations to conformal maps appears to offer advantages. The theory of conformal mapping, with its established existence theorems, is available, and certain invariant properties of conformal mapping might be helpful.

For example, equations, (II.2) may be written in terms of the gradient ∇,

$$\frac{d^2 z_i}{dt^2} = \nabla_{z_i} (V_i + \Phi_i), \qquad i = 2, 3, \cdots, n;$$

where

(II.5)
$$V_i(z_i) = \frac{k^2 M_i}{|z_i|}$$

and

$$\Phi_i (z_2, z_3, \cdots, z_i, \cdots, z_n) = k^2 \sum_{j=2; j \neq i}^{n} \left\{ \frac{m_j}{|z_j - z_i|} - \frac{m_j}{|z_j|} \operatorname{Re}\left(\frac{z_i}{z_j}\right) \right\}.$$

Then, properties of conformal mapping give at once the equivalent of (II.3),

(II.6)
$$\frac{d^2 w_i}{d\tau^2} + \frac{g''(w_i)}{g'(w_i)} \left(\frac{dw_i}{d\tau}\right)^2 - \frac{d^2 t/d\tau^2}{dt/d\tau} \frac{dw_i}{d\tau}$$

$$= \frac{(dt/d\tau)^2}{|g'(w_i)|^2} \nabla_{w_i}(V_i + \Phi_i)', \qquad i = 2, 3, \cdots, n,$$

where $(V_i + \Phi_i)'$ is the transformed $(V_i + \Phi_i)$.

It might be hoped that this behavior of the potential $V_i + \Phi_i$ under conformal mapping is not affected by the interesting fact that this potential is harmonic (statisfies Laplace's equation) in 3-dimensional Euclidean space but is not harmonic in Euclidian 2-space.

III. **Regularization of the relative 3-body equations in the case of binary collisions, using Sundman's transformation.** In this section we will show that the relative 3-body equations can be regularized, in the case of a binary collision, by a transformation due to Sundman of the independent variable only, and the coordinates of the bodies can be expanded in convergent series for all real values of time.

a. *Regularizing Function.* The equations under discussion are obtained from (II.4) as

$$\frac{d^2 z_2}{d\tau^2} - \frac{d^2 t/d\tau^2}{dt/d\tau} \frac{dz_2}{d\tau}$$

(III.1)
$$= \left(\frac{dt}{d\tau}\right)^2 \left\{ - \frac{k^2 M_2}{|z_2|^3} z_2 + k^2 m_3 \left[\frac{z_3 - z_2}{|z_3 - z_2|^3} - \frac{z_3}{|z_3|^3} \right] \right\},$$

$$\frac{d^2 z_3}{d\tau^2} - \frac{d^2 t/d\tau^2}{dt/d\tau} \frac{dz_3}{d\tau}$$

(III.2)
$$= \left(\frac{dt}{d\tau}\right)^2 \left\{ - \frac{k^2 M_3}{|z_3|^3} z_3 + k^2 m_2 \left[\frac{z_2 - z_3}{|z_3 - z_2|^3} - \frac{z_2}{|z_2|^3} \right] \right\}.$$

Denote

$$|z_2| = r_2, \qquad |z_3| = r_3, \qquad |z_3 - z_2| = r_0.$$

Without loss of generality we will consider the binary collision to be that of m_2 with m_3. The Sundman transformation which accomplishes this regularization is given by the equation

(III.3) $\qquad dt = (1 - e^{-r_0/l})(1 - e^{-r_2/l})(1 - e^{-r_3/l})d\tau,$

where l is the positive lower bound of r_2 and r_3, shown by Sundman in [11] to exist. Denote

$$\phi_k = 1 - e^{-r_k/l}, \qquad k = 0, 2, 3.$$

Clearly

$$0 < 1 - \frac{1}{e} \leqq \phi_k < 1, \quad \text{for} \quad r_k < \infty, \qquad k = 2, 3$$

and $0 \leqq \phi_0 < 1$, the equality holding for $r_0 = 0$. Then $0 \leqq dt/d\tau < 1$ and $(dt/d\tau)_{r_0=0} = 0$.

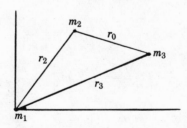

FIGURE 2. Coordinates for Sundman Transformation

Using l'Hôpital's rule we see that

$$\lim_{r_0 \to 0} \frac{dt/d\tau}{r_0} = \lim_{r_0 \to 0} \frac{\phi_2 \phi_3}{l} e^{-r_0/l} = \lim_{r_0 \to 0} \frac{\phi_2 \phi_3}{l} < \frac{1}{l} = m,$$

m a finite quantity,

i.e., $$\frac{dt}{d\tau} = O(r_0),$$

or $dt/d\tau$ vanishes to at least the same order as r_0, as $r_0 \to 0$. Also, in the same notation it follows that

$$\phi_0 = O(r_0)$$

and

$$\phi_k = O(1), \qquad k = 2, 3,$$

i.e.,

$$|\phi_k| \le \text{constant}, r_0 \to 0; \qquad k = 2, 3.$$

Hence the binary collision will not cause a singularity of the differential equations (III.1) and (III.2) provided the following quantities remain finite as $r_0 \to 0$.

$$(\text{III.4}) \qquad \left| \frac{dz_2}{d\tau} \right|, \qquad \left| \frac{dz_3}{d\tau} \right|, \qquad \left| \frac{d^2 t/d\tau^2}{dt/d\tau} \frac{dz_2}{d\tau} \right|, \qquad \left| \frac{d^2 t/d\tau^2}{dt/d\tau} \frac{dz_3}{d\tau} \right|.$$

To show that the limiting values of these quantities are finite, we will make use of the energy integral obtained from the relative 3-body equations of motion in their original form (II.1), before the transformation of the independent variable. These equations are

$$(\text{III.5}) \qquad \frac{d^2 \bar{\rho}_2}{dt^2} = -k^2 \frac{M_2 \bar{\rho}_2}{\rho_2^3} + k^2 m_3 \left[\frac{\bar{\rho}_{23}}{\rho_{23}^3} - \frac{\bar{\rho}_3}{\rho_3^3} \right],$$

$$(\text{III.6}) \qquad \frac{d^2 \bar{\rho}_3}{dt^2} = -k^2 \frac{M_3 \bar{\rho}_3}{\rho_3^3} + k^2 m_2 \left[-\frac{\bar{\rho}_{23}}{\rho_{23}^3} - \frac{\bar{\rho}_2}{\rho_2^3} \right],$$

where

$$M_2 = m_1 + m_2, \qquad M_3 = m_1 + m_3.$$

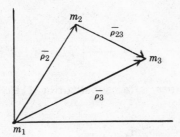

FIGURE 3. Vectors for Energy Integral

Now multiply (III.5) by $m_2\vec{\rho}_2$ and (III.6) by $m_3\vec{\rho}_3$ and add the resulting equations to obtain

$$\sum_{i=2}^{3} m_i \vec{\rho}_i \cdot \vec{\rho}_i = k^2 \left\{ -m_2 m_3 \frac{\vec{\rho}_{23} \cdot \vec{\rho}_{23}}{\rho_{23}^3} - m_2 m_3 \left(\frac{\vec{\rho}_3 \cdot \vec{\rho}_2}{\rho_3^3} + \frac{\vec{\rho}_2 \cdot \vec{\rho}_3}{\rho_2^3} \right) \right.$$

(III.7)

$$\left. - M_2 m_2 \frac{\vec{\rho}_2 \cdot \vec{\rho}_2}{\rho_2^3} - M_3 m_3 \frac{\vec{\rho}_3 \cdot \vec{\rho}_3}{\rho_3^3} \right\} .$$

The term in parentheses on the right side can be written

$$\frac{\vec{\rho}_3 \cdot \vec{\rho}_2}{\rho_3^3} + \frac{\vec{\rho}_2 \cdot \vec{\rho}_3}{\rho_2^3} = \frac{\vec{\rho}_3}{\rho_3^3} \cdot (\vec{\rho}_3 - \vec{\rho}_{23}) + \frac{\vec{\rho}_2}{\rho_2^3} \cdot (\vec{\rho}_2 + \vec{\rho}_{23})$$

(III.8)

$$= \frac{\vec{\rho}_3 \cdot \vec{\rho}_3}{\rho_3^3} + \frac{\vec{\rho}_2 \cdot \vec{\rho}_2}{\rho_2^3} - \vec{\rho}_{23} \cdot \left(\frac{\vec{\rho}_3}{\rho_3^3} - \frac{\vec{\rho}_2}{\rho_2^3} \right).$$

Another expression for the last term of (III.8) is obtained by subtracting (III.5) from (III.6).

$$\ddot{\vec{\rho}}_{23} = \ddot{\vec{\rho}}_3 - \ddot{\vec{\rho}}_2 = k^2 M_2 \frac{\vec{\rho}_2}{\rho_2^3} - k^2 m_3 \frac{\vec{\rho}_{23}}{\rho_{23}^3} + k^2 m_3 \frac{\vec{\rho}_3}{\rho_3^3} - k^2 m_2 \frac{\vec{\rho}_2}{\rho_2^3}$$

$$- k^2 m_2 \frac{\vec{\rho}_{23}}{\rho_{23}^3} - k^2 M_3 \frac{\vec{\rho}_3}{\rho_3^3}$$

$$= k^2 m_1 \frac{\vec{\rho}_2}{\rho_2^3} - k^2 (m_2 + m_3) \frac{\vec{\rho}_{23}}{\rho_{23}^3} - k^2 m_1 \frac{\vec{\rho}_3}{\rho_3^3}$$

$$= -k^2 m_1 \left(\frac{\vec{\rho}_3}{\rho_3^3} - \frac{\vec{\rho}_2}{\rho_2^3} \right) - k^2 (m_2 + m_3) \frac{\vec{\rho}_{23}}{\rho_{23}^3} .$$

Therefore

$$\frac{\vec{\rho}_3}{\rho_3^3} - \frac{\vec{\rho}_2}{\rho_2^3} = \frac{-1}{k^2 m_1} \left[\ddot{\vec{\rho}}_{23} + k^2 (m_2 + m_3) \frac{\vec{\rho}_{23}}{\rho_{23}^3} \right]$$

and

(III.9) $\vec{\rho}_{23} \cdot \left(\dfrac{\vec{\rho}_3}{\rho_3^3} - \dfrac{\vec{\rho}_2}{\rho_2^3} \right) = \dfrac{-1}{k^2 m_1} \left[\vec{\rho}_{23} \cdot \ddot{\vec{\rho}}_{23} + k^2 (m_2 + m_3) \dfrac{\vec{\rho}_{23} \cdot \dot{\vec{\rho}}_{23}}{\rho_{23}^3} \right].$

Then (III.7) becomes, after substituting (III.8) and (III.9) and simplifying,

$$m_2 \dot{\vec{\rho}}_2 \cdot \ddot{\vec{\rho}}_2 + m_3 \dot{\vec{\rho}}_3 \cdot \ddot{\vec{\rho}}_3 + \frac{m_2 m_3}{m_1} \dot{\vec{\rho}}_{23} \cdot \ddot{\vec{\rho}}_{23}$$

$$= - k^2 M \left[m_2 \frac{\vec{\rho}_2 \cdot \dot{\vec{\rho}}_2}{\rho_2^3} + m_3 \frac{\vec{\rho}_3 \cdot \dot{\vec{\rho}}_3}{\rho_3^3} + \frac{m_2 m_3}{m_1} \frac{\vec{\rho}_{23} \cdot \dot{\vec{\rho}}_{23}}{\rho_{23}^3} \right]$$

where $M = m_1 + m_2 + m_3$. Integration gives the energy integral for the relative 3-body equations:

$$\frac{1}{2} (m_1 m_2 \dot{\vec{\rho}}_2^2 + m_1 m_3 \dot{\vec{\rho}}_3^2 + m_2 m_3 \dot{\vec{\rho}}_{23}^2)$$

(III.10)

$$= k^2 M \left(\frac{m_1 m_2}{\rho_2} + \frac{m_1 m_3}{\rho_3} + \frac{m_2 m_3}{\rho_{23}} \right) + E(\text{constant}).$$

From this we see immediately that

$$(m_1 m_2 \dot{\vec{\rho}}_2^2 + m_1 m_3 \dot{\vec{\rho}}_3^2 + m_2 m_3 \dot{\vec{\rho}}_{23}^2) = O\left(\frac{1}{\rho_{23}} \right),$$

i.e., the expression in parentheses goes to infinity to the same order as $1/\rho_{23}$, as $\rho_{23} \to 0$. Hence we have the important results

$$\left| \frac{1}{\dot{\vec{\rho}}_2} \right| = O\left(\frac{1}{\rho_{23}^{1/2}} \right),$$

(III.11) $\left| \dfrac{1}{\dot{\vec{\rho}}_3} \right| = O\left(\dfrac{1}{\rho_{23}^{1/2}} \right),$

$$\left| \frac{1}{\dot{\vec{\rho}}_{23}} \right| = O\left(\frac{1}{\rho_{23}^{1/2}} \right),$$

which imply also

(III.12) $\dot{\rho}_2 = O\left(\dfrac{1}{\rho_{23}^{1/2}} \right), \qquad \dot{\rho}_3 = O\left(\dfrac{1}{\rho_{23}^{1/2}} \right), \qquad \dot{\rho}_{23} = O\left(\dfrac{1}{\rho_{23}^{1/2}} \right).$

In complex notation, the equivalent expressions are

$$|\dot{z}_2| = O\left(\frac{1}{r_0^{1/2}}\right), \qquad |\dot{z}_3| = O\left(\frac{1}{r_0^{1/2}}\right), \qquad |\dot{z}_{23}| = O\left(\frac{1}{r_0^{1/2}}\right)$$

and

$$(\text{III.13})\, \dot{r}_2 = O\left(\frac{1}{r_0^{1/2}}\right), \qquad \dot{r}_3 = O\left(\frac{1}{r_0^{1/2}}\right), \qquad \dot{r}_0 = O\left(\frac{1}{r_0^{1/2}}\right).$$

Returning now to the Sundman transformation (III.3) and the quantities (III.4), we obtain, since $dt/d\tau = O(r_0)$,

$$\left|\frac{dz_2}{d\tau}\right| = \left|\dot{z}_2 \frac{dt}{d\tau}\right| = O(r_0^{1/2})$$

(III.14)

$$\left|\frac{dz_3}{d\tau}\right| = \left|\dot{z}_3 \frac{dt}{d\tau}\right| = O(r_0^{1/2}).$$

Finally, denote

$$\Phi = \frac{dt}{d\tau} = (1 - e^{-r_0/l})(1 - e^{-r_2/l})(1 - e^{-r_3/l})$$

(III.15)

$$= \phi_0 \phi_2 \phi_3.$$

Then

$$\frac{d^2 t}{d\tau^2} = \frac{d\Phi}{d\tau} = \frac{d\Phi}{dt}\frac{dt}{d\tau}$$

and

$$\frac{d^2 t/d\tau^2}{dt/d\tau} = \frac{d\Phi}{dt} = \frac{1}{l}\left\{\phi_0\phi_2\dot{r}_3 e^{-r_3/l} + \phi_0\phi_3\dot{r}_2 e^{-r_2/l} + \phi_2\phi_3\dot{r}_0 e^{-r_0/l}\right\}.$$

Since

$$\phi_0 = O(r_0), \quad \text{and} \quad \phi_k = O(1), \qquad k = 2, 3,$$

expressions (III.13) indicate

$$\frac{d\Phi}{dt} = O\left(\frac{1}{r_0^{1/2}}\right),$$

but using (III.14),

$$\left| \frac{d^2t/d\tau^2}{dt/d\tau} \frac{dz_2}{d\tau} \right| = \left| \frac{d\Phi}{dt} \frac{dz_2}{d\tau} \right| = O(1),$$

and similarly,

$$\left| \frac{d^2t/d\tau^2}{dt/d\tau} \frac{dz_3}{d\tau} \right| = O(1).$$

Thus the Sundman transformation (III.3) or (III.15) does regularize the relative 3-body equations (III.1), (III.2) in the case of a binary collision, and hence these equations are regular for all τ, $|\tau| < \infty$, if we exclude the possibility of triple collision. Therefore the solutions z_2, z_3 are analytic in τ, and can be expressed as power series in $(\tau - \tau_0)$ for arbitrary τ_0, in some interval of convergence.[2] It follows that r_0, r_2 and r_3 are analytic in τ, hence $\Phi = dt/d\tau$ is analytic in τ, $|\tau| < \infty$. Further, since $0 \leqq dt/d\tau < 1$, the equality holding if and only if $r_0 = 0$, then t is a monotone increasing analytic function of τ, $|\tau| < \infty$. Thus there is a one-to-one correspondence between t and τ for t, τ real, $-\infty < t < \infty$, $-\infty < \tau < \infty$.

  b. *Series Solution Convergent for all Finite Values of Time.* If the radius of convergence of the power series in $(\tau - \tau_0)$ for z_2 and z_3 is infinite, the problem is completed, i.e., these series are valid for all real t, because of the one-to-one correspondence between t and τ for all t, τ real.

  Consider therefore that the radius of convergence about τ_0(real) is finite. Since there are no singularities of the differential equations of motion (III.1) and (III.2) for all real τ, the finite radius of convergence about any real τ_0 must be due to the existence of singularities of the solution z_2, z_3 in the finite part of the τ-plane but not in the real axis. Suppose that the singularity which is nearest to the real axis for all τ_0 is at a positive distance h from this axis. (Sundman proves the existence of such a constant h.)

  By the Poincaré transformation

$$\tau - \tau_0 = \frac{2h}{\pi} \log \frac{1+w}{1-w},$$

the interior of the strip of width $2h$ in the τ-plane is transformed

---

[2]Sundman in [11] discusses how the same series define a continuation of the motion after a collision.

conformally onto the interior of the unit circle $|w| = 1$ in the w-plane. This is easily seen through the successive transformations

$$\zeta = e^{(\pi/2h)(\tau - \tau_0)}, \qquad w = \frac{\zeta - 1}{\zeta + 1}.$$

In particular the real axis $-\infty < \tau - \tau_0 < \infty$ is mapped in one-to-one correspondence onto the real segment $-1 < w < 1$.

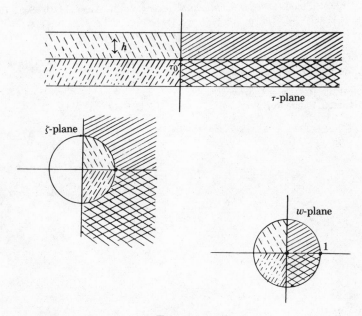

FIGURE 4

The solutions z_2, z_3 are therefore analytic in w, $|w| < 1$, and hence can be expressed in power series in w, convergent in $|w| < 1$. These series will therefore converge for all real w, $|w| < 1$ and consequently for all real τ, $|\tau - \tau_0| < \infty$. Finally, because of the one-to-one correspondence between τ and t for τ, t real, $-\infty < \tau < \infty$, $-\infty < t < \infty$, these series converge for all real time t, $-\infty < t < \infty$.

References

1. E. L. Ince, *Ordinary differential equations,* Dover, New York, 1956.

2. E. A. Coddington and N. Levinson, *Theory of ordinary differential equations,* McGraw-Hill, New York, 1955.

3. S. Lefschetz, *Differential equations: Geometric theory,* Interscience, New York, 1957.

4. V. V. Nemytskiĭ and V. V. Stepanov, *Qualitative theory of differential equations,* Princeton Univ. Press, Princeton, N. J., 1960.

5. F. R. Moulton, *Differential equations,* Dover, New York, 1958.

6. R. P. Agnew, *Differential equations,* McGraw-Hill, New York, 1942.

7. K. Yosida, *Lectures on differential and integral equations,* Interscience, New York, 1960.

8. F. R. Moulton, *Celestial mechanics,* Macmillan, New York, 1914.

9. E. T. Whittaker, *A treatise on the analytical dynamics of particles and rigid bodies,* Cambridge Univ. Press, New York, 1959.

10. G. D. Birkhoff, *Collected mathematical papers,* Vol. 1, Amer. Math. Soc., Providence, R.I. 1950.

11. K. F. Sundman, *Mémoire sur le probléme des trois corps,* Acta Math. **36**(1912), 105.

12. E. Leimanis, *Qualitative methods in general dynamics and celestial mechanics,* Appl. Mech. Rev. **12** No. 10(1959),

AEROSPACE RESEARCH CENTER
GENERAL PRECISION, INC.

John P. Vinti

The Spheroidal Method
in Satellite Astronomy

I. **Introduction.** If $\mathbf{r}$ is the position vector of an artificial satellite of an oblate planet, relative to the latter's center of mass, the drag-free motion of the satellite is determined by the differential equation

(1) $$\ddot{\mathbf{r}} = - \nabla V.$$

Here the gravitational potential V of the planet is expressible as an expansion in spherical harmonics

(2) $$V = -\frac{\mu}{r}\left[1 - \sum_{n=2}^{\infty} \left(\frac{r_e}{r}\right)^n J_n P_n(\sin\theta) \right] + \text{tesseral harmonics}$$

where $r = |\mathbf{r}|$, θ is the declination, r_e is the equatorial radius, P_n is the nth Legendre polynomial, and $\mu = GM$, the product of the gravitational constant and the mass of the planet. Besides the drag, equations (1) and (2) neglect the lunar-solar perturbation and all nongravitational forces. The constants J_n are pure numbers characterizing the planet's potential, with $J_2 = (1.08)10^{-3}$ for the Earth and with all the other J_n's of the order 10^{-6} or smaller.

II. **Possible reference orbits.** Most approaches to the problem of solving (1) and (2) for the orbit have begun with replacing V by $V_0 \equiv - \mu/r$ and finding the perturbations of the resulting elliptic

119

orbit, produced by the higher harmonics. Sterne in [6] and [7] and Garfinkel in [2] and [3] both began with potentials of the form $V = f(r, \theta)$, taking into account part of the effect of the second harmonic. Further progress then depends on finding how the resulting intermediate or reference orbit changes with time.

To take advantage of our knowledge of the actual shape of the earth, or of any oblate planet more closely resembling an oblate spheroid than a sphere, the author[1] in [8] and [9] decided to try oblate spheroidal coordinates. If X, Y, and Z are the usual rectangular coordinates, these spheroidal coordinates ρ, η, and ϕ are defined by the equations

(3.1) $$X + iY = r\cos\theta \exp i\phi = [(\rho^2 + c^2)(1 - \eta^2)]^{1/2}\exp i\phi,$$

(3.2) $$Z = r\sin\theta = \rho\eta.$$

Here c is an adjustable distance, small compared to r_e. For large r, $\rho \to r$ and $\eta \to \sin\theta$. The surfaces $\rho = $ constant are oblate spheroids, approaching sphericity as ρ increases, and the surfaces $\eta = $ constant are hyperboloids, asymptotic to the cones $\theta = $ constant.

With the hope of obtaining a more accurate reference orbit as a starting point for the calculation of satellite orbits, the author wrote out the Hamilton-Jacobi equation in these coordinates, finding that it would be separable if V has the general form

(4) $$V' = (\rho^2 + c^2\eta^2)^{-1}[f(\rho) + g(\eta)].$$

On imposing the requirement that V' shall be a solution of Laplace's equation $\nabla^2 V' = 0$ and that this solution shall be nonsingular on the Z-axis, one finds that the functions $f(\rho)$ and $g(\eta)$ can only be

(5) $$f(\rho) = b_1\rho, \qquad g(\eta) = b_2\eta.$$

Placing the origin at the center of mass then makes $b_2 = 0$ and requiring V' to have the form $-\mu/r$ at large r makes $b_1 = -\mu$. Then

(6) $$V' = -\frac{\mu\rho}{\rho^2 + c^2\eta^2} = -\mu\text{Re}(\rho + ic\eta)^{-1}.$$

---

[1] The author's work in this area has received support from the National Aeronautics and Space Administration.

The expansion of V' in zonal harmonics

$$(7) \qquad V' = -\frac{\mu}{r}\left[1 - \frac{c^2}{r^2}P_2(\sin\theta) + \frac{c^4}{r^4}P_4(\sin\theta) - \frac{c^6}{r^6}P_6(\sin\theta) + \cdots \right]$$

then shows that V' agrees with V through the second harmonic if

$$(8) \qquad c^2 = r_e^2 J_2.$$

With such a choice for c, perfect agreement between V and V' requires that $J_4 = -J_2^2$, $J_6 = J_2^3$, $\cdots$. Since observations show that $J_4 \approx -1.5 J_2^2$, it follows that V' also represents about two-thirds of the fourth harmonic. It follows that V' accounts for about 99.5% of the departure of V from the simple value $-\mu/r$ that would hold for a spherically symmetric planet. In other words the geoid constructed from V' never departs from the actual sea-level surface by more than a few hundred feet. Furthermore, Weinacht in [14] proved that separable motion of a particle in Euclidean space is either a Staeckel system or reducible to a Staeckel system by a point transformation. Of the eleven systems of coordinates in which Staeckel systems may be expressed, the oblate spheroidal has the most appropriate symmetry. Furthermore, equation (6) is the most flexible solution of Laplace's equation in this system that leads to separability. It therefore appears likely that the orbit of a particle moving in the potential field (6), with $c^2 = r_e^2 J_2$, is the best possible reference orbit that can be chosen, from the point of view of accuracy of fit to the actual orbit without perturbation theory.

III. **The quadratures.** If α_1 is the energy, α_3 the axial component of angular momentum, and α_2 a separation constant that would reduce to the total angular momentum in the Keplerian case $c = 0$, then with the potential (6) the Hamilton-Jacobi equation separates, with a solution

$$(9) \qquad W = W_1(\rho, \alpha_1, \alpha_2, \alpha_3) + W_2(\eta, \alpha_1, \alpha_2, \alpha_3) + \alpha_3\phi.$$

If β_1, β_2, and β_3 are constants, such that in the Keplerian case $-\beta_1$ would be the time of passage through perigee, β_2 the argument of perigee, and β_3 the right ascension of the node, the coordinates ρ, η, and ϕ are given by the solution of

$$(10.1) \qquad \frac{\partial W}{\partial \alpha_1} = t + \beta_1 = \pm \int_{\rho_1}^{\rho} \rho^2 F^{-1/2} d\rho \pm c^2 \int_0^{\eta} \eta^2 G^{-1/2} d\eta,$$

$$(10.2) \qquad \frac{\partial W}{\partial \alpha_2} = \beta_2 = \mp \alpha_2 \int_{\rho_1}^{\rho} F^{-1/2} d\rho \pm \alpha_2 \int_0^{\eta} G^{-1/2} d\eta,$$

$$\frac{\partial W}{\partial \alpha_3} = \beta_3 = \phi \mp \alpha_3 \int_0^{\eta} (1 - \eta^2)^{-1} G^{-1/2} d\eta$$

$$(10.3)$$

$$\pm c^2 \alpha_3 \int_{\rho_1}^{\rho} (\rho^2 + c^2)^{-1} F^{-1/2} d\rho.$$

Here

$$(11.1) \qquad \begin{aligned} G(\eta) &= -\alpha_3^2 + (1 - \eta^2)(\alpha_2^2 + 2\alpha_1 c^2 \eta^2) \\ &= (\alpha_2^2 - \alpha_3^2)\left(1 - \frac{\eta^2}{\eta_0^2}\right)\left(1 - \frac{\eta^2}{\eta_2^2}\right) \qquad (-\eta_0 \leqq \eta \leqq \eta_0 \leqq 1) \end{aligned}$$

a quadratic in η^2, and

$$(11.2) \qquad \begin{aligned} F(\rho) &= c^2 \alpha_3^2 + (\rho^2 + c^2)(-\alpha_2^2 + 2\mu\rho + 2\alpha_1 \rho^2) \\ &= (-2\alpha_1)(\rho - \rho_1)(\rho_2 - \rho)(\rho^2 + A\rho + B) \qquad (\rho_1 \leqq \rho \leqq \rho_2) \end{aligned}$$

a quartic in ρ.

IV. **Factoring the quartics.** The material of this and the next two sections is based on [11], [12], and [13].

Finding the coordinates as functions of the time depends on inverting equations (10.1) and (10.2) to obtain ρ and η in terms of t and then inserting the results into (10.3) to obtain ϕ. To do this we must first evaluate the above integrals and this evaluation requires factoring the quartics $F(\rho)$ and $G(\eta)$.

We may define constant orbital elements $a_0 \equiv -\mu/2\alpha_1$, $e_0 \equiv (1 + 2\alpha_1\alpha_2^2/\mu^2)^{1/2}$, and $i_0 \equiv \cos^{-1}(\alpha_3/\alpha_2)$, β_1, β_2, β_3 that can be obtained directly from initial conditions. In this way we can factor $G(\eta)$ exactly and $F(\rho)$ through order J_2^2 without difficulty. A somewhat better set of elements is a, e, I, β_1, β_2, β_3, introduced by Izsak in [4]. Here $a \equiv (\rho_1 + \rho_2)/2$, $e \equiv (\rho_2 - \rho_1)/(\rho_2 + \rho_1)$, and $I = \sin^{-1}\eta_0$. The quantities α_1, α_2, α_3, A, B, and η_2 can all be expressed in terms of these elements, so that the latter lead to an exact factorization of $F(\rho)$. These elements can be obtained from initial conditions by numerical solution of $F(\rho) = 0$, but they can be determined without such a procedure by iterated least-square fitting to an observed orbit.

V. **Evaluating the integrals.** The integrals in (10.3) can be expressed as incomplete elliptic integrals of the third kind and those in (10.1) and (10.2) as incomplete elliptic integrals of the first and second kinds. It is simpler, however, to avoid such a formulation. Suppose we introduce the uniformizing variables E, v, ψ, and χ, defined by

(12.1) $\qquad \rho = a(1 - e\cos E) = a(1 - e^2)(1 + e\cos v)^{-1}$,

(12.2) $\qquad \eta = \eta_0 \sin \psi$,

(12.3) $\quad \exp i\chi = (1 - \eta_0^2 \sin^2 \psi)^{-1/2}(\cos \psi + i(1 - \eta_0^2)^{1/2}\sin \psi)$.

Here E and v are analogous, respectively, to the eccentric and true anomalies in elliptic motion, ψ to the argument of latitude, and χ to the projection of the orbital arc on the equator. By using an expansion in Legendre polynomials with argument $-AB^{-1/2}/2$, viz.,

(13) $\quad (1 + A/\rho + B/\rho^2)^{-1/2} = \sum_{n=0}^{\infty} (B^{1/2}\rho^{-1})^n P_n\left(-\frac{1}{2}AB^{-1/2}\right)$,

we can express the ρ-integrals R_1, R_2, R_3, occurring respectively in (10.1), (10.2), and (10.3), in the forms

(14.1) $\quad (-2\alpha_1)^{1/2}R_1 = -\frac{1}{2}AE + a(E - e\sin E) + A_1 v + \sum_{j=1}^{2} A_{1j}\sin jv$,

(14.2) $\qquad (-2\alpha_1)^{1/2}R_2 = A_2 v + \sum_{j=1}^{4} A_{2j}\sin jv$,

(14.3) $\qquad (-2\alpha_1)^{1/2}R_3 = A_3 v + \sum_{j=1}^{4} A_{3j}\sin jv$.

Here the coefficients A_1, A_2, and A_3 are infinite series, leading to an exact evaluation of secular effects for the reference orbit, and the sine terms are carried far enough to give periodic effects through order J_2^2.

We can express the corresponding η-integrals as

(15.1) $$N_1 = C_1 \psi + \sum_{j=1}^{2} C_{1j} \sin 2j\psi,$$

(15.2) $$N_2 = C_2 \psi + \sum_{j=1}^{2} C_{2j} \sin 2j\psi,$$

(15.3) $$N_3 = C_3 \chi + C_4 \psi + C_{32} \sin 2\psi.$$

Here C_1 and C_2 are expressible in terms of the complete elliptic integrals $K(\eta_0/\eta_2)$ and $E(\eta_0/\eta_2)$ and N_3 in terms of an infinite series. The results for the η-integrals are thus also accurate enough to give secular effects exactly for the reference orbit and periodic effects through order J_2^2.

VI. **Solution of the kinetic equations** (10). One inserts (14) and (15) into equations (10.1) and (10.2), placing

(16) $$E = M_s + E_p, \qquad v = M_s + v_p, \qquad \psi = \psi_s + \psi_p.$$

The secular terms M_s and ψ_s are then found by dropping E_p, v_p, ψ_p and the sine terms in (10.1) and (10.2) and solving a pair of linear algebraic equations. The secular mean anomaly M_s appears as the product of $2\pi\nu_1$ and a linear function of $t + \beta_1$; the secular term ψ_s is the product of $2\pi\nu_2$ and a linear function of $t + \beta_1$. Here ν_1 and ν_2 are, respectively, the mean ρ-frequency $\partial\alpha_1/\partial j_1$ and the mean η-frequency $\partial\alpha_1/\partial j_2$, j_1 and j_2 being the corresponding action variables (see [10]).

One then expresses the periodic terms as

(17) $$E_p = E_0 + E_1 + E_2, \quad v_p = v_0 + v_1 + v_2, \quad \psi_p = \psi_0 + \psi_1 + \psi_2,$$

where E_n, e.g., denotes a periodic part of order J_2^n. One then places $E_p = E_0$, $v_p = v_0$, $\psi_j = \psi_0$ into (10), rejecting all periodic terms of order J_2 or higher, and solves (10.1), (10.2), and (12.1) for E_0, v_0, and ψ_0. Here $M_s + E_0$ appears as the solution of the Kepler equation

(18.1) $$M_s + E_0 - e' \sin(M_s + E_0) = M_s,$$

(18.2) $$e' \equiv ae/a_0 < e.$$

One continues by adding in the terms E_1, v_1, and ψ_1 into equations (10), rejecting only those periodic terms of order J_2^2 or higher. Then

$M_s + E_0 + E_1$ satisfies a similar Kepler equation, the right side getting an additional term M_1, periodic of order J_2, depending on v_0 and $\psi_s + \psi_0$. This second Kepler equation does not require a full-fledged solution, but may be solved by a differential method. Knowing E_1, one may then use (12.1) to find v_1 and (10.2) to find ψ_1.

One continues in a similar fashion to find E_2, v_2, and ψ_2. For the reference orbit the secular parts of E, v, and ψ are then known exactly and the short-periodic parts through order J_2^2. There are no long-periodic terms in this solution for the reference orbit.

Equations (12.1) and (12.2) then give the spheroidal coordinates ρ and η. The right ascension ϕ follows from (10.3), (14.3), and (15.3), after insertion of ψ into (12.3) to find χ. This completes the solution for the reference orbit.

VII. **A sketch of the necessary perturbation theory.** If we subtract (7) from (2), we find that the part of the gravitational potential not accounted for in the reference orbit is given by

$$(19) \quad V - V' = \frac{\mu r_e^3}{r^4} J_3 P_3(\sin\theta) + \frac{\mu r_e^4}{r^5} (J_4 + J_2^2) P_4(\sin\theta) + \cdots.$$

As an example, we consider here only the residual fourth harmonic, so that the perturbing term in the Hamiltonian becomes

$$(20) \qquad\qquad H' = \frac{\mu r_e^4}{r^5} (J_4 + J_2^2) P_4(\sin\theta).$$

If we use the formulas of elliptic motion for r and θ, viz.,

$$(21) \qquad\qquad r = a(1 - e\cos E) = \frac{a(1 - e^2)}{1 + e\cos v},$$

$$(22) \qquad\qquad \sin\theta = \sin I \sin(v + \beta_2)$$

the perturbation H' will be correct through order J_2^2. This order of accuracy will result in secular and short-periodic changes correct through order J_2^2 and long-periodic terms correct through order J_2. (It is well to emphasize at this point that this order of accuracy is for effects produced by less than 0.5% of the departure of the planet from sphericity; for the 99.5% of this departure already accounted for by the potential (7) the secular effects are exact and long-periodic effects do not exist.)

In doing the perturbation theory, the first canonical variables that come to mind are the Jacobi "constants" α_1, α_2, α_3, β_1, β_2, and β_3. Their shortcomings are well known, however, since they lead to Poisson terms in α_1 and α_2. The next set that comes to mind is the one generated from the α's and β's by the generating function

$$(23) \qquad S' = - \alpha_1 t + \mu(- 2\alpha_1)^{-1/2}\beta_1' + \alpha_2\beta_2' + \alpha_3\beta_3'.$$

If we define n_0 by

$$(24) \qquad \mu = n_0^2 a_0^3, \quad a_0 \equiv - \mu/2\alpha_1$$

this leads to the set

$$L = (\mu a_0)^{1/2}, \quad l = n_0(t + \beta_1)$$

$$(25) \qquad\qquad \alpha_2 \qquad\qquad \beta_2$$

$$\qquad\qquad \alpha_3 \qquad\qquad \beta_3,$$

canonical with respect to the Hamiltonian

$$(26) \qquad H = - \mu^2/2L^2 + H'.$$

One may attempt to apply the von Zeipel method in the way successfully used by Brouwer in [1], first eliminating short-periodic terms and then proceeding to eliminate long-periodic terms. One then finds, however, that the corresponding generating function S_1^*, which ought to be of the first order in the parameter

$$(27) \qquad \sigma \equiv J_4 + J_2^2$$

must satisfy

$$(28) \qquad \frac{\partial S_1^*}{\partial \beta_1'} = \text{Zeroth order in } \sigma.$$

One may alternatively eliminate short-periodic and long-periodic terms simultaneously, but one then obtains a Poisson term of the form $v'\sin 2\beta_2'$ in $\alpha_2 - \alpha_2'$. Since v' has a secular part, such a result would appear absurd, since the "constant" α_2, which ought to have only a small periodic variation, would then become infinite. These difficulties are examples of the failure of the von Zeipel method whenever the following conditions *both* hold:

(1) the perturbation has a long-periodic part of the first order in the perturbation parameter σ, and (2) the canonical variables are such that the unperturbed Hamiltonian depends only on L.

The following set, however, is successful. If j_1, j_2, and j_3 are the action variables and if w_1, w_2, and w_3 are the corresponding angle variables, we define L, G, H, l, g, and h by

(29)
$$2\pi L = j_1 + j_2 + j_3 \operatorname{sgn} \alpha_3,$$
$$2\pi G = j_2 + j_3 \operatorname{sgn} \alpha_3,$$
$$2\pi H = j_3,$$

(30)
$$l = 2\pi w_1,$$
$$g = 2\pi (w_2 - w_1),$$
$$h = 2\pi (w_3 - w_2 \operatorname{sgn} \alpha_3),$$

where $\operatorname{sgn} \alpha_3 = \pm 1$ respectively for a direct orbit or a retrograde orbit. To verify that they are canonical, note that

(31)
$$Ldl + Gdg + Hdh = j_1 dw_1 + j_2 dw_2 + j_3 dw_3.$$

They were introduced by Izsak in [5] in his application of the author's theory to the problem of the critical inclination.

If

(32)
$$j_{ik} \equiv \partial j_i / \partial \alpha_k \qquad (i, k = 1, 2, 3)$$

the β's are then given by

(33)
$$2\pi (t + \beta_1) = j_{21} (l + g) + j_{11} l,$$
$$2\pi \beta_2 = j_{22} (l + g) + j_{12} l,$$
$$2\pi \beta_3 = 2\pi h + j_{13} l + (j_{23} + 2\pi \operatorname{sgn} \alpha_3)(l + g).$$

The constant orbital elements in the perturbed problem then become the constant parts a'', e'', and η_0'' of a, e, and η_0, along with the initial values l_0'', g_0'', and h_0'' of the secular parts of l, g, and h.

The corresponding Hamiltonian F is given by

(34)
$$F = F_0(L, G, H) + F_1$$

where

(35)
$$F_0 = -\alpha_1, \qquad F_1 = -H'$$

and

$$\dot{L} = \frac{\partial F}{\partial l}, \qquad \dot{l} = -\frac{\partial F}{\partial L},$$

$$(36) \qquad \dot{G} = \frac{\partial F}{\partial g}, \qquad \dot{g} = -\frac{\partial F}{\partial G},$$

$$\dot{H} = \frac{\partial F}{\partial h}, \qquad \dot{h} = -\frac{\partial F}{\partial H}.$$

One cannot express the unperturbed Hamiltonian $F_0 = -\alpha_1$ exactly as a function of L, G, and H, but it is not necessary to do so. One needs only the derivatives

$$\frac{\partial F_0}{\partial L} = \sum_1^3 \frac{\partial F_0}{\partial j_i} \frac{\partial j_i}{\partial L} = -2\pi\nu_1,$$

$$(37) \qquad \frac{\partial F_0}{\partial G} = \sum_1^3 \frac{\partial F_0}{\partial j_i} \frac{\partial j_i}{\partial G} = 2\pi(\nu_1 - \nu_2),$$

$$\frac{\partial F_0}{\partial H} = \sum_1^3 \frac{\partial F_0}{\partial j_i} \frac{\partial j_i}{\partial H} = 2\pi(\nu_2 \operatorname{sgn}\alpha_3 - \nu_3).$$

On applying the von Zeipel method, one first carries through the simple but tedious elimination of the short-periodic terms. Proceeding to the long-periodic terms, one finds that the appropriate generating function $S_1^*(L', G'', H, g')$ must satisfy

$$(38) \qquad \frac{\partial F_0}{\partial G''} \frac{\partial S_1^*}{\partial g'} = -(F_1)_{\text{long-periodic}},$$

leading to

$$(39) \qquad 2\pi(\nu_1'' - \nu_2'') \frac{\partial S_1^*}{\partial g'} = \sigma f(L', G'', H)\cos 2g',$$

where f is a certain function of L', G'', and H. Since $\nu_1'' - \nu_2''$ is proportional to $1 - 5H^2/G''^2 \approx 1 - 5\cos^2 I$, this leads to the familiar resonance denominator, whenever $\sigma \equiv J_4 + J_2^2 \neq 0$. Since $\sigma/(\nu_1'' - \nu_2'') = O(J_2)$, the long-periodic terms are accurate through order J_2.

After finding the above canonical variables as functions of time, one easily converts their changes into changes of α_1, α_2, and α_3 or of a, e, and I. On inserting the functions $a(t)$, $e(t)$, $I(t)$, $l(t)$, $g(t)$, and $h(t)$ into equations (10), one then can find, by differential methods, the changes in E, v, ψ, and χ, and thus in the coordinates, that are produced by the perturbation. It is not necessary to do a complete re-inversion of (10.1) and (10.2).

References

1. D. Brouwer, *Solution of the problem of artificial satellite theory without drag*, Astronom. J. **64** (1959), 378-397.

2. B. Garfinkel, Astronom. J. **63** (1958), 88-96.

3. _____, *The orbit of a satellite of an oblate planet*, Astronom. J. **64** (1959), 353-367.

4. I. G. Izsak, Smithsonian Inst. Astrophys. Obs., Res. in Space Sci. Special Rep. No. 52, 1960.

5. _____, Smithsonian Inst. Astrophys. Obs., Res. in Space Sci. Special Rep. No. 90, 1962.

6. T. E. Sterne, Astronom. J. **62** (1957), 96.

7. _____, *The gravitational orbit of a satellite of an oblate planet*, Astronom. J. **63** (1958), 28-40.

8. J. P. Vinti, Phys. Rev. Lett. **3** (1959), 98.

9. _____, *New method of solution for unretarded satellite orbits*, J. Res. Nat. Bur. Standards Sect. B **63** (1959), 105-116.

10. _____, *Mean motions in conditionally periodic separable systems*, J. Res. Nat. Bur. Standards Sect. B **65** (1961), 131-135.

11. _____, *Theory of an accurate intermediate orbit for satellite astronomy*, J. Res. Nat. Bur. Standards Sect. B **65** (1961), 169-201.

12. _____, *Formulae for an accurate intermediary orbit for an artificial satellite*, Astronom. J. **66** (1961), 514-516.

13. _____, *Intermediary equatorial orbits of an artificial satellite*, J. Res. Nat. Bur. Standards Sect. B **66** (1962), 5-13.

14. J. Weinacht, Math. Ann. **91** (1924), 279-299.

NATIONAL BUREAU OF STANDARDS

Alan Fletcher

Precession and Nutation

I. **Couple due to distant mass acting on nonspherical object.** Let the axes be principal axes of inertia at the mass-center O (see Figure 1). Let the distant mass be a particle of mass M' at $P(x, y, z)$, and let $r^2 = x^2 + y^2 + z^2$. Let a typical particle of the object be of mass m at $Q(\xi, \eta, \zeta)$, and let

$$\rho^2 = (x - \xi)^2 + (y - \eta)^2 + (z - \zeta)^2.$$

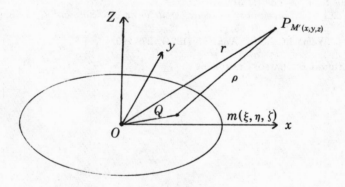

FIGURE 1. Couple Due to Distant Mass

130

Then the force components acting on m are

$$X = GM'm(x - \xi)/\rho^3,$$
$$Y = GM'm(y - \eta)/\rho^3,$$
$$Z = GM'm(z - \zeta)/\rho^3.$$

Thus

$$\widetilde{L} = \text{moment of forces about } Ox$$

$$= \sum (\eta Z - \zeta Y)$$

$$= \sum \frac{GM'm}{\rho^3} [\eta(z - \zeta) - \zeta(y - \eta)]$$

$$= GM' \sum \frac{m(\eta z - \zeta y)}{\rho^3}.$$

But, neglecting squares and products of ξ, η, ζ, we have

$$\rho^2 = (x - \xi)^2 + (y - \eta)^2 + (z - \zeta)^2 = r^2 - 2(x\xi + y\eta + z\zeta),$$

$$\frac{1}{\rho^3} = \frac{1}{r^3} \left[1 + \frac{3(x\xi + y\eta + z\zeta)}{r^2} \right].$$

Therefore, neglecting terms of the third order in (ξ, η, ζ),

$$\widetilde{L} = \frac{GM'}{r^3} \sum m(\eta z - \zeta y) \left[1 + \frac{3(x\xi + y\eta + z\zeta)}{r^2} \right]$$

$$= \frac{3GM'yz}{r^5} \sum m(\eta^2 - \zeta^2),$$

since $\sum m\xi = \sum m\eta = \sum m\zeta = 0$ and $\sum m\eta\zeta = 0$, etc. But if the principal moments of inertia at O are A, B, C, we have

$$A = \sum m(\eta^2 + \zeta^2), \qquad B = \sum m(\zeta^2 + \xi^2), \qquad C = \sum m(\xi^2 + \eta^2),$$

$$\therefore \sum m(\eta^2 - \zeta^2) = C - B,$$

(1) $\quad \therefore \widetilde{L} = -\dfrac{3GM'(B - C)yz}{r^5}, \qquad \widetilde{M} = -\dfrac{3GM'(C - A)zx}{r^5},$

$$\widetilde{N} = -\frac{3GM'(A - B)xy}{r^5}.$$

II. **Case of an axially symmetrical body (e.g. the Earth).** In this case $A = B < C$, and the couple has components

(2) $$\frac{3GM'(C-A)yz}{r^5}, \qquad -\frac{3GM'(C-A)zx}{r^5}, \qquad 0.$$

If $x = r\cos\delta$, $y = 0$, $z = r\sin\delta$, so that δ is the declination of the distant object, the couple components are

(3) $$0, \qquad -\frac{3GM'(C-A)}{r^3}\sin\delta\cos\delta, \qquad 0,$$

i.e., the couple is in the plane of M' and the Earth's axis, and is in such a sense that (if the Earth were not rotating) it would tend to move the Earth's equatorial plane into coincidence with M'.

III. **Couple components derived from potential energy.** Let U be the gravitational potential at P due to the nonspherical object, so that the mutual potential energy of this and the mass M' at P is $V = -M'U$. To the order of approximation already assumed in §I, U is given by MacCullagh's formula, proved in many text books,

$$U = \frac{GM}{r} + \frac{G(A+B+C-3I)}{2r^3},$$

where I is the moment of inertia of the nonspherical object about OP and M is its mass. Thus

(4) $$V = -\frac{GMM'}{r} - \frac{GM'(A+B+C-3I)}{2r^3}.$$

From this the general formulas (1) for $\widetilde{L}$, $\widetilde{M}$, $\widetilde{N}$ may easily be deduced by considering changes in V produced by small rotations about the axes of coordinates.

Restricting ourselves to the case of axial symmetry $(A = B)$, we have

$$I = A\cos^2\delta + C\sin^2\delta,$$

where δ is the declination of P. Thus

$$A + B + C - 3I = 2A + C - 3(A\cos^2\delta + C\sin^2\delta)$$
$$= (C-A)(1 - 3\sin^2\delta).$$

Therefore

(5) $$V = -\frac{GMM'}{r} - \frac{GM'(C-A)(1-3\sin^2\delta)}{2r^3},$$

where it may be noted that

$$\frac{1}{2}(3\sin^2\delta - 1) = P_2(\sin\delta),$$

P_2 being the Legendre polynomial of the second degree. Thus the couple due to M' tending to decrease δ has a magnitude

$$\frac{\partial V}{\partial \delta} = \frac{3GM'(C-A)}{r^3}\sin\delta\cos\delta$$

which agrees with (3).

This method is useful if it is required to ascertain the possible effect of the fourth harmonic. Taking

$$U = \frac{GM}{a}\left[\frac{a}{r} + J\frac{a^3}{r^3}\left(\frac{1}{3} - \sin^2\delta\right) + \frac{8}{35}D\frac{a^5}{r^5}P_4(\sin\delta)\right]$$
$$= U_0 + U_2 + U_4,$$

we have

$$P_4(\sin\delta) = \frac{1}{8}(35\sin^4\delta - 30\sin^2\delta + 3)$$

and

$$\frac{\partial P_4}{\partial \delta} = \frac{1}{2}(35\sin^2\delta - 15)\sin\delta\cos\delta,$$

which is numerically less than or equal to $10\sin\delta\cos\delta$. Thus

$$\left|\frac{\partial U_4/\partial\delta}{\partial U_2/\partial\delta}\right| \leq \frac{\frac{8}{35}D \times 10}{J \times 2}\left(\frac{a}{r}\right)^2 = \frac{8D}{7J}\left(\frac{a}{r}\right)^2.$$

With $J = 1.6 \times 10^{-3}$, $D = 8 \times 10^{-6}$, so that D/J is about $1/200$, and with $a/r = 1/60$ in the case when M' is the Moon, this gives about $1/600,000$. Since we shall not determine precessions to more than 4 significant figures, the effect of the fourth harmonic in the

Earth's potential may be neglected.

IV. **Average couple when M' moves in a circle around the Earth.**
Suppose M' moves, for example, in the plane of the ecliptic (see
Figure 2). Let P and K denote the poles of the equator and the

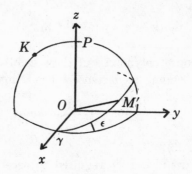

FIGURE 2. Mass in the Ecliptic

ecliptic respectively. Take the x-axis along $O\gamma$. If

$$\kappa = \frac{3GM'(C-A)}{2r^3},$$

which in turn is equal to the maximum numerical value of the couple,
occurring at $\delta = \pm 45°$, the couple has, by (2), the components
$2\kappa yz/r^2$, $-2\kappa zx/r^2$, 0. But $x = r\cos L$, $y = r\sin L\cos\epsilon$, $z = r\sin L\sin\epsilon$,
where ϵ is the obliquity of the ecliptic and $L = \text{angle }\gamma OM'$ increases
uniformly with the time. Thus

$$yz = r^2\sin\epsilon\cos\epsilon\sin^2 L \quad \text{and} \quad zx = r^2\sin\epsilon\sin L\cos L$$

have average values $(1/2)\,r^2\sin\epsilon\cos\epsilon$ and 0 respectively. Thus the
average couple has components $(\widetilde{N}', 0, 0)$, where

$$\widetilde{N}' = \kappa\sin\epsilon\cos\epsilon = \frac{3GM'(C-A)}{2r^3}\sin\epsilon\cos\epsilon.$$

If the Earth were not rotating, this couple would tend to move
the pole P of the equator towards the pole K of the ecliptic.

We shall now consider the effect of such couples on the Earth,
supposing it rigid. (The Earth is deformable, but not enough to
affect the precession seriously, as has been found by several authors,
and as is obvious from consideration of angular momentum.) The

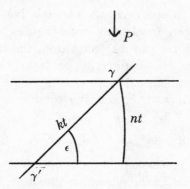

FIGURE 3. Rotation of the Equator

following is an approximate treatment, which ignores for the moment the distinction between the Earth's axis of figure and axis of rotation.

V. **Effect of steady couple on the rotating Earth.** Suppose a steady couple $\widetilde{N}'$ about $O\gamma$ acts on the Earth (axial moment of inertia C, angular velocity ω). In a small time t, the angular momentum $C\omega$ about OP receives a vector increment $\widetilde{N}'t$ about $O\gamma$. Thus P moves towards the instantaneous position of γ at rate $n = \widetilde{N}'/C\omega$, and KP rotates about K at a rate $k = n \operatorname{cosec} \epsilon$.

$$(6) \qquad n = \text{motion of pole} = \frac{3GM'}{2r^3\omega} \frac{C-A}{C} \sin\epsilon \cos\epsilon,$$

$$(7) \qquad k = \text{motion of equinox} = \frac{3GM'}{2r^3\omega} \frac{C-A}{C} \cos\epsilon.$$

Here $(C-A)/C$, the Earth's "mechanical ellipticity", is found (actually from the precession) to be about $1/(305.5)$.

Neglecting the eccentricity of the Earth's orbit and the eccentricity, inclination, and solar perturbations of the Moon's orbit, the above formulas give the solar and lunar precessions, on substituting solar and lunar values of M' and r. (Variability of the couples produces nutation, to be considered later.)

VI. **Rough numerical values.** For the Sun, $GM' = n'^2 r^3$, thus k from the Sun per year $(2\pi/n')$ is

$$3\pi \frac{n'}{\omega} \frac{C-A}{C} \cos\epsilon = \frac{1944000'' \cos 23°\,27'}{366.24 \times 305.5} = 15''.9,$$

which is correct to one decimal. The relative importance of the Sun and the Moon (neglecting eccentricities, etc.) in producing precession (also tides) may be found from the ratio of their values of M'/r^3. This is found, directly, or from $(M/(E+M))(n/n')^2$ (see §XII), to be about 2.17. Thus k from the Moon per year is about $2.17 \times 15''.9 = 34''.5$, and the total luni-solar precession is about

$$k = 15''.9 + 34''.5 = 50''.4/\text{year} = 1°.4/\text{century}.$$

VII. **Luni-solar precession.** If the ecliptic were fixed, P would go around K in a circle, in the retrograde direction, with period $360/1.4$ centuries, or about 25,700 years, with constant inclination ϵ. The longitude of a fixed star would increase by about $50''/\text{year}$, its latitude remaining constant. That is, $\Delta\lambda = kt$, $\Delta\beta = 0$, or $\lambda = k$, $\beta = 0$. With $\epsilon = 23°27'$, $\sin\epsilon \doteq 2/5$, so that motion of pole $= n$ $= k\sin\epsilon \doteq 20''/\text{year}$. The effect of luni-solar precession (only) on right ascension and declination is shown in books on spherical astronomy (e.g. [13]) to be given by

$$\alpha = n \cot\epsilon + n \sin\alpha \tan\delta,$$

$$\delta = \qquad n \cos\alpha.$$

VIII. **Effect of motion of the ecliptic (planetary precession).** The plane of the ecliptic slowly rotates on account of planetary perturbations of the Earth's orbit. Let N, N' be the ascending and descending nodes of the ecliptic at time $t + dt (dt > 0)$ on the ecliptic at time t; let $\gamma N = 180° - \nu$ and $N'\gamma = \nu$. At 1950.0, $\nu = 5°36'$, and $j =$ speed of rotation of ecliptic $= 0''.471/\text{year}$, which is less than $1/40$ of the polar motion n.

Resolve the rotation of the plane of the ecliptic into

(i) $j\cos\nu$ about the line of equinoxes, decreasing ϵ, so that $\dot\epsilon$ $= -j\cos\nu$. At 1950.0, $\dot\epsilon = -0''.469/\text{year}$ (see Figure 5).

(ii) $j\sin\nu$ about the line of solstices, giving $\gamma\gamma'\sin\epsilon = j\sin\nu$; thus γ would move *forwards* on a fixed equator at a rate $j\sin\nu/\sin\epsilon$ $= l$ (see Figure 4). Thus l (which is slowly decreasing) is the planetary precession (in R.A.), decreasing all right ascensions. At 1950.0, $j\sin\nu = 0''.046$, $l = j\sin\nu/\sin\epsilon = 0''.115$ (both yearly).

IX. **Full formulas for $\dot\alpha$, $\dot\delta$.** Planetary precession contributes $\dot\alpha = -l$, $\dot\delta = 0$ (since declinations are not affected). Thus

FIGURE 4. Rotation of the Ecliptic

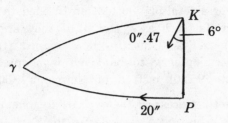

FIGURE 5. Motions of the Poles of the Equator and the Ecliptic

$$\dot{\alpha} = m + n \sin \alpha \tan \delta,$$

$$\dot{\delta} = \quad\quad n \cos \alpha,$$

where $m =$ "precession in R.A." $= n \cot \epsilon - l = k \cos \epsilon - l$. General precession (in longitude) is accordingly

$$p = n \operatorname{cosec} \epsilon - l \cos \epsilon = k - l \cos \epsilon.$$

The formulas are often used as $\Delta\alpha = (\bar{m} + \bar{n} \sin \bar{\alpha} \tan \bar{\delta}) \Delta t,$ $\Delta\delta = \bar{n} \cos \bar{\alpha} \Delta t,$ where $\bar{m},$ $\bar{n}$ refer to mid-epoch and $\bar{\alpha},$ $\bar{\delta}$ either refer to mid-epoch or are means of initial and final values of $\alpha,$ $\delta.$ But rigorous formulas must be used for long intervals or large $\delta.$

X. **Numerical values for** 1950.0.

$$\epsilon = 23°27' \qquad n = 20''.04 \qquad k = n \operatorname{cosec} \epsilon = 50''.37$$

$$m = n \cot \epsilon - l = 46''.10 \qquad p = k - l \cos \epsilon = 50''.27.$$

The above are Newcomb's values, rounded to 4 figures. They take the mean of the stars to be at rest. But two factors affect the fourth figure, in the ways discussed in the next section.

XI. **Effect of galactic rotation and relativity.** (i) It is now known that the mean of the stars is not at rest, because the galaxy is rotating. The effect on proper motions is of the order of $1''$ per century. See, for example, [9], [10], [3], [14, p. 382], [8] and [7, p. 16].

(ii) The dynamical interpretation of the precession also needs modification. The Newtonian dynamics being given really applies in a frame rotating at $1''.94$ per century *directly* around the pole of the ecliptic. See, for example, [3] and [12].

The effect of (i) and (ii) is that corrections of about $+0''.01$ and $+0''.02$, or a total of $+0''.03$, are needed by annual precessions in longitude (such as k and p) if they are to be used to determine $(C-A)/C$. Thus the more refined treatment of the luni-solar precession now to be given is adapted to the value $k = 50''.40$ (at 1950.0) instead of Newcomb's $50''.37$.

XII. **Luni-solar precession (more exact treatment).** If n, a refer to the Moon's orbit and n', a' refer to the Sun's orbit, let

$$K_1 = \frac{3GS}{2a'^3\omega}\frac{C-A}{C}, \qquad K_2 = \frac{3GM}{2a^3\omega}\frac{C-A}{C}.$$

Here S and M are the masses of the Sun and Moon, and K_1, K_2 are constants such that (by the theory of §V) the Sun and the Moon would cause polar motions

$$n_1 = K_1\cos\epsilon\sin\epsilon, \quad n_2 = K_2\cos\epsilon\sin\epsilon$$

and precessions in longitude

$$k_1 = K_1\cos\epsilon, \quad k_2 = K_2\cos\epsilon$$

if the eccentricity e' of the Earth's orbit and the eccentricity e and inclination i of the Moon's orbit were zero and the Moon's motion were not perturbed by the Sun. Let $E =$ the mass of the Earth. Then

$$G(S+E) = n'^2a'^3, \qquad G(E+M) = n^2a^3.$$

Thus

$$K_1 = \frac{3}{2}\frac{S}{S+E}\frac{n'^2}{\omega}\frac{C-A}{C}, \qquad K_2 = \frac{3}{2}\frac{M}{E+M}\frac{n^2}{\omega}\frac{C-A}{C}.$$

$E/S = 3\times10^{-6}$ and may be neglected here. Put $\mu = M/E$. Then

$$K_1 = \frac{3}{2}\frac{n'^2}{\omega}\frac{C-A}{C}, \qquad K_2 = \frac{3}{2}\frac{\mu}{1+\mu}\frac{n^2}{\omega}\frac{C-A}{C}.$$

Now

$$\frac{3}{2}\frac{n'^2}{\omega} = \frac{3}{2}\frac{n'}{\omega/n'} = \frac{3}{2}\frac{1296000''/\text{year}}{366.24} = 5308''/\text{year}.$$

Using $(C - A)/C = 1/305.5$ we find that the value of K_1 is

$$K_1 = 17''.37 \text{ per year}.$$

Also

$$\frac{K_2}{K_1} = \frac{M/a^3}{S/a'^3} = \frac{\mu}{1+\mu}\left(\frac{n}{n'}\right)^2.$$

Using $\mu = 1/81.3$ and $n'/n = 0.07480$ we find that

$$K_2/K_1 = 2.172.$$

As $2.172 \times 17.37 = 37.73$, we find that the value of K_2 is

$$K_2 = 37''.73 \text{ per year}.$$

XIII. **Allowance for eccentricities, etc.** To allow for eccentricity of orbits, we require the average of $1/r^3$ rather than $1/a^3$. But it is easily shown that in a Keplerian orbit the average of r^{-3} is $a^{-3}(1 - e^2)^{-3/2}$, so that the factor $(1 - e^2)^{-3/2}$, or approximately $1 + (3e^2/2)$, allows for eccentricity.

In addition, for the Moon, if a is defined by $G(E + M) = n^2a^3$, a is *not* the mean distance. The variational orbit gives

$$a/r = 1 + \frac{1}{6}m^2 + m^2\cos 2\xi + O(m^3),$$

where $m = n'/n$ temporarily, and ξ varies linearly with the time. Thus

$$(a/r)^3 = 1 + \frac{1}{2}m^2 + 3m^2\cos 2\xi + O(m^3),$$

and the average of $(a/r)^3$ is $1 + \frac{1}{2}m^2$ approximately.

Thus, omitting squares and products of e'^2, e^2, m^2, define

(8) $$P_1 = \left(1 + \frac{3}{2}e'^2\right)K_1,$$

(9) $$P_2 = \left(1 + \frac{3}{2}e^2 + \frac{1}{2}m^2\right)K_2.$$

This takes approximate account of everything but the inclination of the Moon's orbit to the ecliptic. The m^2 correction is often omitted in textbooks, but is of the same order as the e^2 correction. A much more complete orbital correction is given in [2]. With $e' = 0.0167$, $e = 0.05490$, $m = 0.07480$, we find

(10) $P_1 = 1.00042 K_1 = 17''.38$ per year,

(11) $P_2 = (1 + 0.00452 + 0.00280) K_2 = 1.00732 K_2 = 38''.01$ per year.

P_2 now needs correction for the inclination $i = 5°9'$. We shall prove in working out nutation (see §XXII) that the factor is $1 - (3/2) \sin^2 i$ approximately. Thus let

(12)
$$P_2' = \left(1 + \frac{3}{2} e^2 - \frac{3}{2} \sin^2 i + \frac{1}{2} m^2\right) K_2$$
$$= (1.00732 - 0.0121) K_2 = 0.9952 K_2,$$

or rather we shall take the more accurate value

(13) $P_2' = 0.99537 K_2 = 37''.56$ per year.

Then

$$P = P_1 + P_2' = 54''.94 \text{ per year.}$$

This is the most constant quantity connected with precession, and Newcomb proposed to call it the precessional constant. He gave its value as $54''.9066$ per year, decreasing only $0''.0000364$ per century. The difference of about $0''.03$ has been explained in §XI. Spencer Jones, in ([7], p. 16), gives

$$P = (54''.93553 \pm 0''.00145) \text{ per year.}$$

Note that ϵ is slowly decreasing,
$n = P \cos\epsilon \sin\epsilon$ is slowly decreasing,
$k = P \cos\epsilon$ is slowly increasing,
$m = P \cos^2\epsilon - l$ is slowly increasing, partly because ϵ is decreasing and partly because l is decreasing.

At 1950.0 we have (for $\mu = 1/81.3$, $(C - A)/C = 1/(305.5)$)

(14)
$P_1 \cos\epsilon = 15''.94$ per year,

$P_2 \cos\epsilon = 34''.87$ per year,

$P_2' \cos\epsilon = 34''.46$ per year,

$P \cos\epsilon = 50''.40$ per year (compare $50''.37$ Newcomb).

XIV. **Axis of figure and axis of rotation.** A polhode and herpolhode argument shows that the difference in direction in steady precession is about

$$\frac{23.5 \times 3600''}{25700 \times 366} = 0''.009.$$

XV. **Nutation.** The motion of the pole of the equator may be divided into a secular part (precession) and periodic parts (nutation).

If at any time K is the pole of the ecliptic, P_0 is the mean pole of the equator and P is the true pole of the equator (see Figure 6), we define the nutations as follows:

(1) nutation in obliquity $= \Delta\epsilon = KP - KP_0$,

(2) nutation in longitude $= \Delta\psi = P_0KP$, positive as shown; note that $\Delta\psi$ also equals $\gamma_0\gamma$, positive as shown, and is equal to $\Delta\lambda$, the true minus mean longitude of any star. Nutation arises from various causes, and its complete determination is rather complicated. Much the largest part arises from the retrogression of the Moon's nodes with period about 18.6 years; we shall investigate this, and consider briefly two other effects.

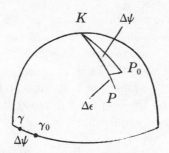

FIGURE 6. Motion of Pole of Equator

The final results of complete investigations show that the largest terms are

$$\Delta\epsilon = 9''.21\cos\Omega - 0''.09\cos2\Omega + 0''.55\cos2L + 0''.09\cos2\mathbb{C}\,,$$

$$\Delta\psi = -17''.23\sin\Omega + 0''.21\sin2\Omega - 1''.27\sin2L - 0''.20\sin2\mathbb{C}\,,$$

where the largest neglected coefficients are about $0''.02$ in $\Delta\epsilon$ and $0''.13$ in $\Delta\psi$. Here

Ω = longitude of ascending node of Moon's orbit on ecliptic,

L = Sun's mean longitude,

$\mathbb{C}$ = Moon's mean longitude.

XVI. Orbit of the Moon. The period relative to either the fixed stars or the equinox is 27.322 days. The orbit is inclined at an average inclination of $5°9'$ ($\pm 9'$ on either side) to the ecliptic, and the nodes regress (not quite uniformly) in a period 18.60 years relative to the fixed stars, or 18.61 years relative to the equinox.

XVII. Precessional effects of Sun and Moon. The Sun and Moon, according to the theory of §XIII, if moving at any angle ϵ to the equator, would set up an average motion of pole $n = (P_1 + P_2) \sin \epsilon \cos \epsilon$ and precession in longitude $k = (P_1 + P_2) \cos \epsilon$, where $P_1(\text{Sun}) = 17''.38$ per year and $P_2(\text{Moon}) = 38''.01$ per year are practically constant. At 1950.0 we have

$$P_1 \cos \epsilon = 15''.94 \text{ per year}, \quad P_2 \cos \epsilon = 34''.87 \text{ per year}.$$

In investigating nutation, we shall also deal with the correction of P_2 for inclination in evaluating precession.

XVIII. Effect of Moon's orbital inclination and nodal retrogression. In Figure 7, K is the pole of the ecliptic, M the pole of the Moon's orbit, P the pole of the equator. M moves backwards around K in a circle of radius $i = 5°9'$. Thus P describes a wavy path. Let $\theta = PKM = 180° - \Omega$, where Ω = longitude of the ascending node of the Moon's orbit on the ecliptic. θ *increases* by 2π in 18.61 years. If $\Omega = N_0 - N_1 t$, then approximately

$$\dot{\theta} = N_1 = \frac{2\pi}{18.61} = 0.3376 \text{ rad/year}.$$

As 18.61 years is long compared with a month, we may suppose that the Moon is giving P a velocity $P_2 \cos PM \sin PM$ perpendicular to PM. If ξ and η are the small displacements of P along and perpendicular to KP, we have

$$\dot{\xi} = P_2 \cos PM \sin PM \sin KPM,$$

$$\dot{\eta} = P_2 \cos PM \sin PM \cos KPM.$$

ξ is the nutation in obliquity, $\Delta\epsilon$; η gives a secular part (lunar precession) and a periodic part ($\Delta\psi \sin \epsilon$).

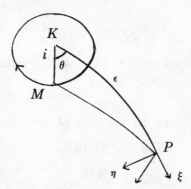

FIGURE 7. Motion of Pole of Moon's Orbit

XIX. **Nutation in obliquity.** Using $\sin PM \sin KPM = \sin i \sin \theta$, we have

$$\dot{\xi} = P_2 \sin i \sin \theta \cos PM$$

$$= P_2 \sin i \sin \theta (\cos \epsilon \cos i + \sin \epsilon \sin i \cos \theta)$$

$$= \frac{1}{2} P_2 \cos \epsilon \sin 2i \sin \theta + \frac{1}{2} P_2 \sin \epsilon \sin^2 i \sin 2\theta.$$

But $\dot{\theta} = N_1$, so that integration gives

$$\xi = -\frac{P_2 \cos \epsilon \sin 2i}{2N_1} \cos \theta - \frac{P_2 \sin \epsilon \sin^2 i}{4N_1} \cos 2\theta$$

$$= -N \cos(180° - \Omega) - \frac{1}{4} N \tan \epsilon \tan i \cos(360° - 2\Omega);$$

or

$$\Delta\epsilon = N \cos \Omega - \frac{1}{4} N \tan \epsilon \tan i \cos 2\Omega,$$

where

$$N = \text{constant of nutation}$$

$$= \frac{P_2 \cos \epsilon \sin 2i}{2N_1} = \frac{34''.87 \sin 10°17'26''}{2 \times 0.3376} = 9''.226,$$

and

$$\frac{1}{4} N \tan \epsilon \tan i = 0''.09.$$

Thus

$$\Delta\epsilon = 9''.23\cos\Omega - 0''.09\cos 2\Omega,$$

and this explains the major part of the nutation in obliquity due to the rotation of the Moon's orbit. The observed coefficient of $\cos\Omega$ is $9''.21$.

XX. Polar motion in the direction of decreasing longitude.

$$\dot\eta = P_2\cos PM\sin PM\cos KPM$$

$$= P_2\cos PM\sin PM\frac{\cos i - \cos\epsilon\cos PM}{\sin\epsilon\sin PM}$$

$$= P_2(\cos i\operatorname{cosec}\epsilon\cos PM - \cot\epsilon\cos^2 PM).$$

So

$$\dot\eta = P_2\cos i\operatorname{cosec}\epsilon\,(\cos\epsilon\cos i + \sin\epsilon\sin i\cos\theta)$$

$$- P_2\cot\epsilon\,(\cos\epsilon\cos i + \sin\,\epsilon\sin i\cos\theta)^2.$$

Using $\cos^2\theta = (1 + \cos 2\theta)/2$, $\dot\eta$ may be split up into a constant part (giving lunar precession) and periodic parts (giving nutation).

XXI. Periodic parts.

$$\dot\eta = P_2\cos i\sin i\,(1 - 2\cos^2\epsilon)\cos\theta - \frac{1}{2}\,P_2\cos\epsilon\sin\epsilon\sin^2 i\cos 2\theta$$

$$= -\frac{1}{2}\,P_2\cos 2\epsilon\sin 2i\cos\theta - \frac{1}{4}\,P_2\sin 2\epsilon\sin^2 i\cos 2\theta.$$

Thus integration gives

$$\eta = -\frac{P_2\cos 2\epsilon\sin 2i}{2N_1}\,\sin\theta - \frac{P_2\sin 2\epsilon\sin^2 i}{8N_1}\,\sin 2\theta$$

$$= -N\frac{\cos 2\epsilon}{\cos\epsilon}\,\sin\Omega + \frac{1}{4}\,N\sin\epsilon\tan i\sin 2\Omega$$

$$= -6''.87\sin\Omega + 0''.08\sin 2\Omega$$

(using $N = 9''.226$). Using the observed $N = 9''.21$, the coefficient of $\sin\Omega$ is $-6''.86$. Thus, neglecting the small terms in 2Ω, the observed N gives

$$\xi = 9''.21\cos\Omega, \qquad \eta = -6''.86\sin\Omega,$$

so that the equatorial pole P describes an ellipse (shown in Figure 8) with major axis about $18''.4$ along $P_0 K$ and minor axis about $13''.7$, around the mean pole P_0, with period 18.61 years, in the sense indicated (remembering that Ω decreases).

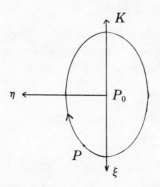

FIGURE 8. Polar Nutation.

The nutation in longitude is given by

$$\Delta\psi \sin\epsilon = \text{the above periodic part of } \eta,$$

so that

$$\Delta\psi = -2N\cot 2\epsilon \sin\Omega + \frac{1}{4}N\tan i \sin 2\Omega$$

$$= -17''.27 \sin\Omega + 0''.21 \sin 2\Omega$$

(using $N = 9''.226$). Using $N = 9''.21$ gives $17''.24$. This explains the major part of the nutation in longitude due to the rotation of the Moon's orbit.

XXII. **Constant part (lunar precession).** This should reduce to $P_2 \cos\epsilon \sin\epsilon$ when $i = 0$. The constant part is given by

$$\eta = P_2 \cot\epsilon \cos^2 i - P_2 \cot\epsilon \left(\cos^2\epsilon \cos^2 i + \frac{1}{2}\sin^2\epsilon \sin^2 i \right)$$

$$= P_2 \cos\epsilon \sin\epsilon \left(1 - \frac{3}{2}\sin^2 i \right) = P_2 \cos\epsilon \sin\epsilon \frac{1 + 3\cos 2i}{4}$$

$$= P_2' \cos\epsilon \sin\epsilon,$$

where

$$P_2' = P_2 \left(1 - \frac{3}{2}\sin^2 i\right) = 0.98794\,P_2 = 0.98794 \times 38''.01$$

$$= 37''.55 \text{ per year.}$$

The factor $1 - (3/2)\sin^2 i$ is needed in dealing with the lunar part of the precession (see §XIII).

XXIII. **Nutation due to variability of the solar couple during the year.** The average solar couple $(\kappa \sin\epsilon \cos\epsilon, 0, 0)$ of §IV results from averaging the actual couple

$$(2\kappa \sin^2 L \sin\epsilon \cos\epsilon, \ -2\kappa \sin L \cos L \sin\epsilon, \ 0),$$

where we are supposing the Sun's orbit circular and using L for the Sun's longitude and

$$\kappa = 3GS(C - A)/2a'^3.$$

In discussing precession, we have taken account of the average couple; we have now to take account of the periodic couple

$$(-\kappa \sin\epsilon \cos\epsilon \cos 2L, \ -\kappa \sin\epsilon \sin 2L)$$

about Ox and Oy of Figure 2 respectively. Putting $K_1 = \kappa/C\omega$, these produce polar motions respectively

$-K_1\sin\epsilon \cos\epsilon \cos 2L$ along $P\gamma$ and $-K_1\sin\epsilon \sin 2L$ along KP.

But in the notation of §XVIII, these are $\dot{\eta}$ and $\dot{\xi}$ respectively, i.e.,

$$\dot{\xi} = -K_1\sin\epsilon \sin 2L,$$

$$\dot{\eta} = -\frac{1}{2} K_1\sin 2\epsilon \cos 2L.$$

If $\dot{L} = n' = 2\pi/\text{year}$, we have on integrating

$$\xi = \frac{K_1\sin\epsilon}{2n'} \cos 2L = \Delta\epsilon, \qquad \eta = -\frac{K_1\sin 2\epsilon}{4n'} \sin 2L = \Delta\psi \sin\epsilon,$$

$$\Delta\psi = -\frac{K_1\cos\epsilon}{2n'} \sin 2L.$$

Using $K_1 = 17''.37/\text{year}$, $n' = 2\pi/\text{year}$, we have

$$\Delta\epsilon = \xi = 0''.55 \cos 2L,$$

$$\Delta\psi \sin\epsilon = \eta = -0''.50 \sin 2L,$$

$$\Delta\psi = \eta \operatorname{cosec}\epsilon = -1''.27 \sin 2L,$$

the coefficient of $\sin 2L$ in $-\Delta\psi$ being simply $1/2\pi \times$ solar precession in longitude in 6 months, or $7''.97/2\pi$.

This explains the six-monthly terms due to the Sun.

XXIV. **Nutation due to variability of lunar couple during the month.** Neglecting the inclination of the lunar orbit to the ecliptic, we have quite similar results for the Moon. If $n = 2\pi/27^d.322 = 13.37 \times 2\pi/\text{year}$, we have

$$\Delta\epsilon = \frac{K_2 \sin \epsilon}{2n} \cos 2\mathbb{C} = \frac{37''.73 \sin 23°27'}{13.37 \times 4\pi} \cos 2\mathbb{C} = 0''.09 \cos 2\mathbb{C},$$

$$\Delta\psi = -\frac{K_2 \cos \epsilon}{2n} \sin 2\mathbb{C} = -\frac{37''.73 \cos 23°27'}{13.37 \times 4\pi} \sin 2\mathbb{C} = -0''.21 \sin 2\mathbb{C},$$

the coefficient of $\sin 2\mathbb{C}$ in $-\Delta\psi$ being simply $1/2\pi \times$ lunar precession in longitude in $(1/2) \times 27.322$ days. This explains the fortnightly terms due to the Moon.

We have now derived all the terms given in §XV.

XXV. **Difference between axis of figure and axis of rotation.** If the orientation of the Earth is specified by Eulerian angles, and these have been determined as functions of the time, the position of the axis of rotation at any time may easily be deduced (from Euler's geometrical equations). Woolard in ([16], p. 159) finds that no nutation coefficient in ξ or η differs by more than $0''.0062$ according to which axis is used; there is also a constant difference of $0''.0087$ in ξ arising from precession (see §XIV); the axes may, however, differ in direction by up to $0''.4$ on account of variation of latitude. Federov in [1] uses the axis of angular momentum, which never differs by more than about $0''.001$ from the axis of rotation.

The results in Table 1, to three decimals of a second of arc, appear to follow for 1900.0 from ([16], pp. 153 and 159); see also ([6], p. 78).

XXVI. **Comparison of theory and observations.** The usual method has been to solve for the nutation constant N, all other nutation coefficients being regarded as known multiples of N. It appears from ([5], p. 224) and [6] that the difference of one or two hundredths of a second of arc between the calculated (rigid body) and the observed values of N may be accounted for geophysically (some

Table 1. Coefficients for Nutation

$$\xi = \Delta\epsilon \ \text{(obliquity)}$$

	$\cos\Omega$	$\cos 2\Omega$	$\cos 2L$	$\cos 2\mathbb{C}$
Rotation	$+9''.210$	$-0''.090$	$+0''.552$	$+0''.088$
Figure	$+9\ .209$	$-0\ .090$	$+0\ .555$	$+0\ .094$
Difference	$+0\ .001$	0	$-0\ .003$	$-0\ .006$

$$\eta = \Delta\psi \sin\epsilon$$

	$\sin\Omega$	$\sin 2\Omega$	$\sin 2L$	$\sin 2\mathbb{C}$
Rotation	$-6''.858$	$+0''.083$	$-0''.507$	$-0''.081$
Figure	$-6\ .857$	$+0\ .083$	$-0\ .510$	$-0\ .087$
Difference	$-0\ .001$	0	$+0\ .003$	$+0\ .006$

$$\eta \csc\epsilon = \Delta\psi \ \text{(longitude)}$$

	$\sin\Omega$	$\sin 2\Omega$	$\sin 2L$	$\sin 2\mathbb{C}$
Rotation	$-17''.233$	$+0''.209$	$-1''.273$	$-0''.204$
Figure	$-17\ .230$	$+0\ .209$	$-1\ .280$	$-0\ .220$
Difference	$-0\ .003$	0	$+0\ .007$	$+0\ .016$

Earth models would allow a greater difference, and the actual difference is well within the bounds of possible explanation). For astronomical purposes, it seems sufficient to adopt the observed value of N.

Federov in [1] (see also [6]) has determined the chief coefficients independently, and considered phases (by determining a cosine coefficient when rigid body theory predicts a sine term, and vice versa). He appears to find no clear difference from the usual ratios and phases which is of much importance for practical astronomical purposes.

References

1. E. P. Federov, *Nutatsiya i vynuzhdennoe dvizhenie polyusov zemli...*, Poltavskaya Gravimetr. Obs., Izdat. Akad. Nauk Ukrainskoi SSR, Kiev, 1958.
2. G. W. Hill, Astronom. J. **13** (1893), 1-6 (Coll. Math. Works **4** (1907), 11-21; Carnegie Institution).

3. J. Jackson, Monthly Notices Roy. Astr. Soc. **90** (1930), 733-742.

4. H. Jeffreys, *Dynamics of the Earth-moon system*, The Earth as a planet, (= The solar system. II, G. P. Kuiper, ed.), Univ. Chicago Press, Chicago, Ill., 1954; Chapter II.

5. _____ , *The earth*, 4th ed. Cambridge Univ. Press, New York, 1959. 1959.

6. _____ , Monthly Notices Roy. Astr. Soc. **119** (1959), 75-80.

7. H. Spencer Jones, *Dimensions and rotation*, The Earth as a planet, (= The solar system. II, G. P. Kuiper, ed.), Univ. Chicago Press, Chicago, Ill., 1954; Chapter I.

8. H. R. Morgan, Bull. Astronom. **15** (1950), 199-211.

9. J. H. Oort, Bull. Astronom. Inst. Netherlands (132) **4** (1927), 79-89.

10. _____ , Bull. Astronom. **15** (1950), 217-228.

11. H. C. Plummer, *An introductory treatise on dynamical astronomy*, Cambridge Univ. Press, New York, 1918.

12. W. deSitter and D. Brouwer, Bull. Astronom. Inst. Netherlands (307) **8** (1938), 213-231.

13. W. M. Smart, *Textbook on spherical astronomy*, 1st ed., Cambridge Univ. Press, New York, 1931.

14. _____ , *Stellar dynamics*, Cambridge Univ. Press, New York, 1938. 1938.

15. _____ , *Celestial mechanics*, Longmans, Green, London, 1953.

16. E. W. Woolard, *Theory of the rotation of the Earth around its center of mass*, Astr. Pap. Amer. Ephem. (1) **15** (1953), 165 pp.

UNIVERSITY OF LIVERPOOL

Victor Szebehely

On an Irreversible Dynamical System
with Two Degrees of Freedom:
the Restricted Problem
of Three Bodies

Introduction. The three series of lectures, which I delivered at the Yale Dynamical Astronomy Institutes of 1961, 1962 and 1963 respectively on the restricted problem, dealt with three different faces of this celebrated problem in dynamics.

In the 1961 series of lectures I set a rather ambitious goal and followed a *qualitative approach*. Birkhoff's and Poincaré's fundamental contributions formed the basis of these lectures, which dealt with periodic motions, reducibility, regularization, regions of possible motions, etc.

The 1962 lecture series concentrated on what might be termed the *formalistic treatment* of dynamical problems. The powerful tools of celestial mechanics, canonical transformations and variables were used and the pertinent equations of motion were derived. Performing regularizations within the framework of canonical transformations requires the introduction of the concept of the extended phase space. Once the method of canonical transformations is extended to this $2n + 2 = 6$ dimensional space, a great variety of forms of the equations of the restricted problem are obtainable. As examples, Cartesian rectangular coordinates (synodic and sidereal), polar coordinates, Delaunay variables, Poincaré's variables, etc. were derived.

The third series of lectures, delivered in 1963 at Cornell University under the auspices of the American Mathematical Society, presented the *quantitative results* of the restricted problem. The numerical work of the famous Copenhagen school was used as a basis with which G. Darwin's, F. Moulton's, and V. Egorov's results were compared. A large number of recent contributions were reviewed and the existence of certain families of orbits at various mass ratios was established.

The trilogy described above followed the three approaches to dynamics. Because of the different nature of these investigations, their results are also substantially different and repetitions are easily avoidable. Nevertheless, all three are designed to gain a better understanding of the same nonintegrable, irreversible, two-degree-of-freedom dynamical system.

The vast amount of material covered in the three lecture series is not suitable for condensation in a short article. A thorough treatise of the restricted problem with its applications will shortly appear in my book, *Theory of orbits—The restricted problem of three bodies,* Academic Press, New York (to appear in 1966). However, the present article should be useful to those readers who wish to become familiar with the basic ideas of the restricted problem and who wish to see its relation to the general problem of three bodies, to regularization, etc. A few typical results of the formalistic and qualitative approaches will be sketched. The considerable collection of references will enable the reader to pursue the subject further.

Statement of the problem and equations of motion. The problem under discussion is the restricted problem of three bodies (problème restreint). Two bodies (assumed to be point masses and called primaries) revolve around their center of mass in circular orbits under the influence of their mutual gravitational attraction. A third body (which is attracted by the previous two but is not influencing their motion) moves in the plane defined by the two revolving bodies. The problem is to determine the motion of this third body.

Let the masses of the two primary bodies be m_1 and m_2, their mean motion (angular velocity) n, and their distance l (see Figure 1, in which $l = a + b$). Then

(1)
$$k^2 M = n^2 l^3$$

where $M = m_1 + m_2$.

The center of mass of the system is located on the line connecting m_1 and m_2, and its distances from m_2 and m_1 are respectively

$$(2) \qquad a = \frac{m_1 l}{M}; \quad b = \frac{m_2 l}{M}.$$

Taking the origin, 0, of a fixed inertial coordinate system (X, Y) at the mass center, and using τ for time, the equations of motion referred to this system will be

$$(3) \qquad \frac{d^2 X}{d\tau^2} = \frac{\partial F}{\partial X}, \quad \frac{d^2 Y}{d\tau^2} = \frac{\partial F}{\partial Y},$$

where F is Poincaré's "force function" or the negative potential energy and it is given by

$$(4) \qquad F = k^2 \left(\frac{m_1}{\rho_1} + \frac{m_2}{\rho_2} \right)$$

with ρ_1, and ρ_2 being the distances between the primaries and the third body.

We note that the Hamiltonian,

$$(5) \qquad H = \frac{1}{2} \left[\left(\frac{dX}{d\tau} \right)^2 + \left(\frac{dY}{d\tau} \right)^2 \right] - F$$

is not constant, since F depends explicitly on the time,

$$F = F(x, Y, \tau).$$

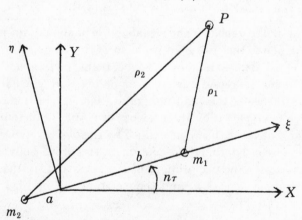

FIGURE 1. The Restricted Problem in Fixed (X, Y) and Rotating (ξ, η) Coordinate Systems

Introducing a uniformly rotating coordinate system (ξ, η) with origin at the mass center, so that m_1 and m_2 are located on the ξ axis with coordinates $(b, 0)$ and $(-a, 0)$, the equations of motion become

(6)
$$\frac{d^2\xi}{d\tau^2} - 2n\frac{d\eta}{d\tau} = \frac{\partial F^*}{\partial \xi},$$
$$\frac{d^2\eta}{d\tau^2} + 2n\frac{d\xi}{d\tau} = \frac{\partial F^*}{\partial \eta},$$

where

(7)
$$F^* = F + \frac{1}{2}n^2(\xi^2 + \eta^2)$$

and

(8)
$$\rho_1^2 = (\xi - b)^2 + \eta^2, \qquad \rho_2^2 = (\xi + a)^2 = \eta^2.$$

The introduction of nondimensional quantities simplifies the equations. Let

$$x = \xi/l, \quad y = \eta/l, \quad r_1 = \rho_1/l, \quad r_2 = \rho_2/l,$$

$$t = n\tau, \quad \mu = \frac{m_2}{M}, \quad \text{and} \quad \overline{\Omega} = F^*/l^2 n^2.$$

The equations of motion in nondimensional form are

(9)
$$\frac{d^2x}{dt^2} - 2\frac{dy}{dt} = \frac{\partial \overline{\Omega}}{\partial x},$$
$$\frac{d^2y}{dt^2} + 2\frac{dx}{dt} = \frac{\partial \overline{\Omega}}{\partial y},$$

and we have the relations

(10)
$$\overline{\Omega} = \frac{1}{2}(x^2 + y^2) + \frac{\mu}{r_2} + \frac{1-\mu}{r_1}$$

and

(11)
$$r_1^2 = (x - \mu)^2 + y^2, \qquad r_2^2 = (x + 1 - \mu)^2 + y^2.$$

Equations (9), (10) and (11) represent the problem in conventional nondimensional quantities. The corresponding physical picture is as follows: the two primary bodies are located on the x axis which rotates with unit angular velocity; the coordinates of the primaries are $P_1\,(\mu, 0)$ and $P_2\,(\mu - 1, 0)$, and their masses are $1 - \mu$

and μ, with $0 \leq \mu \leq 1$; the distance between the primaries, their total mass, their angular velocity and the gravitational constant are unity.

The Jacobi integral is obtained by multiplying the first of equations (9) by $2dx/dt$ and the second by $2dy/dt$, adding and integrating:

$$(12) \qquad \left(\frac{dx}{dt}\right)^2 + \left(\frac{dy}{dt}\right)^2 = 2\overline{\Omega} - \overline{C},$$

where $\overline{C}$ is the constant of integration.

We note that according to Equations (9) the partial derivatives of $\overline{\Omega}$ completely determine the problem, therefore adding a constant to the expression for $\overline{\Omega}$—as given by Equation (10)—will not change Equations (9), but will influence the value of $\overline{C}$ in Equation (12). A symmetrical form of $\overline{\Omega}$ is obtained by adding the constant quantity $\mu(1 - \mu)/2$ to $\overline{\Omega}$. Let $\Omega = \overline{\Omega} + \frac{1}{2}\mu(1 - \mu)$, or

$$(13) \qquad \Omega = \frac{1}{2}[(1 - \mu)r_1^2 + \mu r_2^2] + \frac{1 - \mu}{r_1} + \frac{\mu}{r_2}.$$

The new Jacobi constant C is related to the previous one by

$$C = \overline{C} + \mu(1 - \mu).$$

The final set of equations, with the notations

$$\Omega_x = \frac{\partial \Omega}{\partial x}, \quad \Omega_y = \frac{\partial \Omega}{\partial y}, \quad \dot{x} = \frac{dx}{dt}, \quad \text{etc.,}$$

becomes:

$$(14) \qquad \ddot{x} - 2\dot{y} = \Omega_x, \qquad \ddot{y} + 2\dot{x} = \Omega_y,$$

$$(15) \qquad (\dot{x})^2 + (\dot{y})^2 = 2\Omega - C,$$

where Ω is given by Equation (13).

Relationship to the general problem of three bodies. The general problem of three bodies is defined as follows: three particles attract each other according to the Newtonian law of gravitation, they are free to move in space and are initially moving in any given manner; find their subsequent motion. Their masses are m_1, m_2, and m_3, and their position vectors from the origin of coordinates are $\overline{r_1}$, $\overline{r_2}$, and $\overline{r_3}$ (see Figure 2).

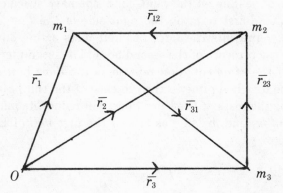

FIGURE 2. The General Problem of Three Bodies

The potential function of the system is

(16) $$V = -k^2 \left(\frac{m_1 m_2}{|\bar{r}_{12}|} + \frac{m_2 m_3}{|\bar{r}_{23}|} + \frac{m_3 m_1}{|\bar{r}_{31}|} \right).$$

The equations of motion can be written as:

(17) $$m_i \ddot{\bar{r}}_i = -\frac{\partial V}{\partial \bar{r}_i} \qquad i = 1, 2, 3,$$

where the right-hand sides represent the gradients of the potential in the directions of the position vectors. These equations can also be written as:

(18) $$\ddot{\bar{r}}_1 = -k^2 m_2 \frac{\bar{r}_1 - \bar{r}_2}{|\bar{r}_1 - \bar{r}_2|^3} + k^2 m_3 \frac{\bar{r}_3 - \bar{r}_1}{|\bar{r}_3 - \bar{r}_1|^3},$$

$$\ddot{\bar{r}}_2 = -k^2 m_3 \frac{\bar{r}_2 - \bar{r}_3}{|\bar{r}_2 - \bar{r}_3|^3} + k^2 m_1 \frac{\bar{r}_1 - \bar{r}_2}{|\bar{r}_1 - \bar{r}_2|^3},$$

$$\ddot{\bar{r}}_3 = -k^2 m_1 \frac{\bar{r}_3 - \bar{r}_1}{|\bar{r}_3 - \bar{r}_1|^3} + k^2 m_2 \frac{\bar{r}_2 - \bar{r}_3}{|\bar{r}_2 - \bar{r}_3|^3}.$$

Equations (18) describe the general case of the problem of three bodies with Newtonian gravitational forces. The structure of these equations is of some interest inasmuch as the masses m_1, m_2, and m_3 are missing from the first, second, and third equations respectively. This fact does not "uncouple" the equations since all three position vectors occur in all three equations.

The terms appearing on the right-hand side have specific physical significance. The first term on the right side of the first equation, for instance, represents the force per unit mass acting on the first body due to the presence of the second body. The second term on the right side of the same equation represents the force per unit mass acting on the first body due to the presence of the third body.

Decreasing the mass of the third body will reduce its influence on the motion of m_1 and m_2, i.e., as $m_3 \rightarrow 0$ the first two of Equations (18) become

(19)
$$\ddot{r}_1 = - k^2 m_2 \frac{\overline{r}_1 - \overline{r}_2}{|\overline{r}_1 - \overline{r}_2|^3},$$

$$\ddot{r}_2 = - k^2 m_1 \frac{\overline{r}_2 - \overline{r}_1}{|\overline{r}_2 - \overline{r}_1|^3},$$

while the third equation will not change. This step does uncouple the equations since the motion of m_1 and of m_2 can now be determined without considering the effect of the third mass by solving the 12th order system of differential equations (19).

The third equation of (18) requires comment since if in the first two equations $m_3 = 0$ (which gives Equations (19)) then the third equation should be $0 = 0$. This is because the usual derivation of the third equation of (18) from the third equation of (17) by dividing through by m_3 is not permitted when $m_3 = 0$. Continuing to use the third equation of (18) constitutes the approximation which creates the restricted problem of three bodies. In effect, the assumption is made that $m_3 \neq 0$ but that m_3 is sufficiently small so that it does not effect the motion of m_1 and of m_2. That is, Equations (19) are approximate while the third equation of (18) is exact. The system of equations consisting of Equations (19) and of the third equation of (18) represent the true dynamical system only approximately; the degree of approximation is given by the "smallness" of the terms

$$m_3 \frac{\overline{r}_3 - \overline{r}_1}{|\overline{r}_3 - \overline{r}_1|^3} \quad \text{and} \quad m_3 \frac{\overline{r}_2 - \overline{r}_3}{|\overline{r}_2 - \overline{r}_3|^3}$$

as compared to the terms

$$m_2 \frac{\overline{r}_1 - \overline{r}_2}{|\overline{r}_1 - \overline{r}_2|^3} \quad \text{and} \quad m_1 \frac{\overline{r}_1 - \overline{r}_2}{|\overline{r}_1 - \overline{r}_2|^3}$$

respectively.

The effects of the first two bodies on the motion of the third is given by the third equation of (18). Accepting the above-mentioned approximation, one might solve Equations (19), substitute the solutions into the third equation of (18), and obtain the sixth order differential equation for the motion of the third body:

$$(20) \qquad \ddot{r_3} = -k^2 m_1 \frac{\overline{r_3} - \overline{r_1}(t)}{|\overline{r_3} - \overline{r_1}(t)|^3} + k^2 m_2 \frac{\overline{r_2}(t) - \overline{r_3}}{|\overline{r_2}(t) - \overline{r_3}|^3},$$

where now $\overline{r_1}(t)$ and $\overline{r_2}(t)$ are given functions of the time and of the initial conditions, and $\overline{r_3}$ is to be determined as a function of time.

Equation (20) describes the restricted problem of three bodies. The "restriction" is equivalent to the assumption according to which the motion of the first two bodies is not influenced by the third, while the motion of the third is determined by the masses and by the motions of the first two.

Equation (20) can be generalized to

$$(21) \qquad \ddot{r_3} = \overline{f}(m_1, m_2, \overline{r_1}(t), \overline{r_2}(t), \overline{r_3}),$$

where $\overline{r_3}$ is the only unknown function of time, m_1, and m_2 are given constants and $\overline{r_1}(t)$ and $\overline{r_2}(t)$ are also given as functions of the time and of their initial conditions.

The function $\overline{f}$ in Equation (21) represents the force field which in Equation (20) was the Newtonian gravitational field. In addition to Equation (21) the initial conditions of the third body will be needed, in order to determine its motion.

The classification of the various forms of the restricted problem follows from the above remarks and is based on Equation (21).

1. Depending on the force law, $\overline{f}$, we speak of Newtonian and non-Newtonian problems.

2. Depending on the initial conditions for $\overline{r_1}$ and $\overline{r_2}$, in the Newtonian case, we speak of the circular or of the general-conic-section restricted problems.

3. Depending on the initial conditions of the third body we distinguish between the planar and the three-dimensional restricted problems. In a Newtonian gravitational field the first two bodies (primaries or principal bodies) will always move in a plane. If the third body's initial velocity vector is in the plane determined by the orbits of the primaries, it will stay in this plane; this is the case in which Equations (14) and (15) are applicable.

4. Further classification is possible by specifying the m_2/m_1 ratio. The Copenhagen restricted problem, for instance, is distinguished by using unity for this ratio.

Reductions of the order of differential equations of motion. The fourth order system of Equations (14) can be reduced to a third order system by making use of the Jacobi integral, Equation (15). One way to accomplish this is by introducing the angle z between the tangent to the orbit and the positive x axis as a new dependent variable:

$$(22) \qquad\qquad z = \arctan \frac{dy}{dx}.$$

We will show that the equations of motion become:

$$(23) \qquad \begin{aligned} \dot{x} &= \Lambda(x, y)\cos z, \\ \dot{y} &= \Lambda(x, y)\sin z, \\ \dot{z} &= -2 - \Lambda_x \sin z + \Lambda_y \cos z, \end{aligned}$$

where

$$(24) \qquad\qquad \Lambda = (2\Omega - C)^{1/2}.$$

The proof of the first two equations of (23) is as follows. If s denotes the arc length,

$$\frac{dx}{ds} = \cos z, \qquad \frac{dy}{ds} = \sin z,$$

and so

$$\dot{x} = \frac{dx}{ds}\dot{s} = \Lambda \cos z, \qquad \dot{y} = \Lambda \sin z,$$

since the absolute value of the velocity vector $\dot{s}$ is obtained from the Jacobi integral as

$$\left(\frac{ds}{dt}\right)^2 = (\dot{x})^2 + (\dot{y})^2 = 2\Omega - C = \Lambda^2.$$

The proof of the third equation of (23) requires the evaluation of $\dot{z}$ from Equation (22):

$$\dot{z} = \frac{\ddot{y}\dot{x} - \dot{y}\ddot{x}}{(\dot{s})^2}.$$

Using Equations (14) to eliminate $\ddot{x}$ and $\ddot{y}$, and the first two equations of (23) to eliminate $\dot{x}$ and $\dot{y}$, we obtain the desired result.

We note that an alternate approach would be to use the first two equations of (23) as definitions of a transformation without reference to Equation (22). In this way we have $\dot{x} = \Lambda \cos z, \dot{y} = \Lambda \sin z$ and consequently $\ddot{x} = \dot{\Lambda} \cos z - \Lambda \dot{z} \sin z$ and $\ddot{y} = \dot{\Lambda} \sin z + \Lambda \dot{z} \cos z$, from which $\Lambda \dot{z}$ can be obtained; $\Lambda \dot{z} = \ddot{y} \cos a - \ddot{x} \sin z$. By the same elimination process as before we can obtain the third equation of (23).

Equations (23) represent the third-order version of the problem in the form of three first-order differential equations with x, y, z as dependent and t as the independent variable. The significant fact is noted that these equations can be written as

$$
\begin{aligned}
\dot{x} &= \phi(x, y, z), \\
(25) \qquad \dot{y} &= \psi(x, y, z), \\
\dot{z} &= \chi(x, y, z).
\end{aligned}
$$

That is, the right-hand members do not contain the time. This fact allows a physical interpretation of the equations by an analogy and also assures further reduction of the order by elimination of the time.

Consider a flow field with velocity vector $\bar{v} = \bar{v}(\bar{r}, t)$, where $\bar{r}$ is the position vector and t is the time. This velocity vector gives a description of the flow field since at every point $\bar{r}$ in the field, at any time t, the velocity can be evaluated—excepting singular points. A flow field is called steady if

$$
\frac{\partial \bar{v}}{\partial t} = 0,
$$

i.e., if none of the velocity components depend explicitly on the time. The velocity components defined by Equations (25) can be interpreted therefore as the description of a steady flow field.

The continuity equation of hydrodynamics is

$$
\frac{\partial \rho}{\partial t} + \text{div}(\rho \bar{v}) = 0.
$$

where ρ is the fluid density. For an incompressible fluid $\rho = \text{constant}$ and the continuity equation becomes

$$
\text{div} \, \bar{v} = 0,
$$

or using the notation of Equations (25)

$$\phi_x + \psi_y + \chi_z = 0.$$

Since the ϕ, ψ, χ velocity components, as given by Equations (23) satisfy this equation, the dynamical problem is analogous to the three-dimensional steady flow of an incompressible fluid.

It should be remarked that the flow is *not* a potential flow since

$$\operatorname{curl} \bar{v} \neq 0;$$

direct computation of the six partial derivatives $\phi_y, \phi_z, \psi_x, \psi_z, \chi_x, \chi_y$ from Equations (23) will establish this.

The function Λ contains C, therefore Equations (23) will determine a flow field for each given C value. By changing the constant of integration, C, the streamline picture will change. For a given C, Equations (23) will determine the totality of motions of the dynamical system and also the corresponding streamlines, providing that the inequality.

$$\left(\frac{ds}{dt}\right)^2 = 2\Omega(x, y) - C \geqq 0$$

is satisfied.

The streamline representation is singular when $\Lambda = 0$ or when $\Lambda \to \infty$. The first case corresponds to zero velocity, the second to collisions at P_1 or P_2.

Poincaré's original flow analogy is similar to but not identical in detail with the one just given, and should be mentioned also. Let $x = x_1, y = x_2, \dot{x} = x_3,$ and $\dot{y} = x_4$. Then the equations of motion can be written as

$$\dot{x}_1 = x_3,$$

$$\dot{x}_2 = x_4,$$

$$\dot{x}_3 = 2x_4 + \frac{\partial\Omega(x_1, x_2)}{\partial x_1},$$

$$\dot{x}_4 = -2x_3 + \frac{\partial\Omega(x_1, x_2)}{\partial x_2},$$

or

$$\dot{x}_i = F_i(x_1, x_2, x_3, x_4), \qquad i = 1, \cdots, 4.$$

It can be seen that $\operatorname{div} \bar{v} = 0$, i.e.,

$$\sum_{i=1}^{4} \frac{\partial F_i}{\partial x_i} = 0,$$

so we are dealing with the four-dimensional, steady stream-line flow of an incompressible fluid. The actual motion of the particle corresponds to those streamlines which lie on the

$$x_3^2 + x_4^2 - 2\Omega(x_1, x_2) + C = 0$$

three-dimensional hypersurface.

Equations (25) can be written as

(26) $$\frac{dx}{\phi} = \frac{dy}{\psi} = \frac{dz}{\chi},$$

from which

(27) $$\frac{dz}{dx} = \frac{\chi}{\phi}.$$

On the other hand Equation (22) gives

(28) $$\frac{dz}{dx} = \frac{d}{dx} \arctan y' = \frac{y''}{1 + (y')^2},$$

where $y' = dy/dx$.

Substituting for χ and ϕ in Equation (27) their expressions as given by Equations (23), eliminating z by Equation (22) and equating the right sides of Equations (27) and (28) results in a second-order differential equation describing the dynamical system (excepting at points where the transformations are singular):

(29) $$y'' = \frac{1 + (y')^2}{\Lambda} \left[\Lambda_y - \Lambda_x y' \pm 2(1 + (y')^2)^{1/2} \right].$$

It should be noted that elimination of time and use of the Jacobi integral can be combined and the above second-order differential equation can be obtained in a single step directly from the original fourth-order system. The general solution of (29) will contain the following three constants of integration: C which is included in Λ and the two constants which enter when (29) is integrated. The complete solution of the original fourth-order system requires the determination of the time dependence of the variables. This process will result in the fourth integration constant. To establish the time dependence, we write the Jacobi integral as

$$(\dot{x})^2[1 + (y')^2] = \Lambda^2,$$

where the $\dot{y} = \dot{x}y'$ relation was used. The time is evaluated from the last equation as

$$t = \int \frac{(1 + (y')^2)^{1/2}}{\Lambda} \, dx + C_4.$$

Here y' and $\Lambda(x,y)$ are functions of x only since $y(x)$ has been obtained from Equation (29).

Regularization of the restricted problem. The main purpose of this section is to show how regularization is performed within the framework of canonical variables utilizing the concept of the extended phase space. We will conduct the discussion on a generalized level so that the reader can use this section as his starting point for the solution of new problems.

The treatment starts with the Hamiltonian in a rotating rectangular Cartesian coordinate system:

$$(30) \qquad H = \frac{1}{2}(p_1^2 + p_2^2) + q_2 p_1 - q_1 p_2 - F(q_1, q_2),$$

where q_1, q_2 are uniformly rotating Cartesian rectangular coordinates (corresponding to x and y of Equation (14)), and p_1, p_2 are the conjugate momenta $(p_1 = \dot{x} - y, \ p_2 = \dot{y} + x)$. We also recall that the gravitational part of Ω is

$$F = \frac{1 - \mu}{r_1} + \frac{\mu}{r_2},$$

where

$$(31) \qquad r_1^2 = (q_1 - \mu)^2 + q_2^2, \quad \text{and} \quad r_2^2 = (q_1 + 1 - \mu)^2 + q_2^2.$$

The generating function to be used is a so-called extended point transformation:

$$(32) \qquad W_3 = p_1 f(Q_1, Q_2) + p_2 g(Q_1, Q_2),$$

where f and g are conjugate harmonic functions of the new coordinates Q_1 and Q_2.

The corresponding transformation equations are

$$(33) \qquad q_i = \frac{\partial W_3}{\partial p_i} \quad \text{and} \quad P_i = \frac{\partial W_3}{\partial Q_i}, \qquad i = 1, 2.$$

First the coordinate tranformation is performed in the phase space then the transformation in the extended phase space is executed.

Equations (32) and (33) give

$$p_1^2 + p_2^2 = \frac{1}{D}(P_1^2 + P_2^2),$$

where

$$D(Q_1, Q_2) = \left(\frac{\partial f}{\partial Q_1}\right)^2 + \left(\frac{\partial g}{\partial Q_1}\right)^2.$$

The new Hamiltonian also requires the computation of the $q_2 p_1 - p_2 q_1$ term, which is obtained by substitution:

$$\frac{1}{2D}\left[P_1 \frac{\partial}{\partial Q_2}(f^2 + g^2) - P_2 \frac{\partial}{\partial Q_1}(f^2 + g^2)\right].$$

Therefore the new Hamiltonian becomes

$$(34) \quad \widetilde{H} = \frac{1}{2D}\left[P_1^2 + P_2^2 + \left(P_1 \frac{\partial}{\partial Q_2} - P_2 \frac{\partial}{\partial Q_1}\right)(g^2 + f^2) - \widetilde{F}(Q_1, Q_2)\right],$$

where in computing $\widetilde{F}$ we write f in place of q_1 and g in place of q_2 in Equations (31).

The equations of motion are

$$(35)$$
$$\dot{Q}_1 = \frac{1}{2D}\left[2P_1 + \frac{\partial}{\partial Q_2}(f^2 + g^2)\right],$$
$$\dot{Q}_2 = \frac{1}{2D}\left[2P_2 - \frac{\partial}{\partial Q_1}(f^2 + g^2)\right],$$

$$(36)$$
$$\dot{P}_1 = -\frac{\partial \widetilde{H}}{\partial Q_1},$$
$$\dot{P}_2 = -\frac{\partial \widetilde{H}}{\partial Q_2}.$$

Inspecting the last term of $\widetilde{H}$ we realize that Equations (36) have singularities at the primaries.

In the extended phase space we have the Hamiltonian

$$(37) \quad \Gamma = P_3 + \frac{1}{2D}\left[P_1^2 + P_2^2 + \left(P_1 \frac{\partial}{\partial Q_2} - P_2 \frac{\partial}{\partial Q_1}\right)(f^2 + g^2)\right]$$
$$- \widetilde{F}(Q_1, Q_2),$$

and the equations of motion are reducible to Equations (35) and (36).

We now introduce $\Gamma^* = D\Gamma$ and obtain

$$(38) \quad \Gamma^* = DP_3 + \frac{1}{2}\left[P_1^2 + P_2^2 + \left(P_1\frac{\partial}{\partial Q_2} - P_2\frac{\partial}{\partial Q_1}\right)(f^2 + g^2)\right] - D\widetilde{F},$$

and with this the equations of motion using the new time variable $(\bar{t})$ can be established. The relation between dt and $d\bar{t}$ is $dt = Dd\bar{t}$ since $\Gamma^* = D\Gamma$. Denoting derivatives with respect to $d\bar{t}$ by primes, we have

$$Q_i' = \frac{\partial \Gamma^*}{\partial P_i}, \quad P_i' = -\frac{\partial \Gamma^*}{\partial Q_i}, \quad i = 1, 2, 3,$$

or

$$(39) \quad
\begin{aligned}
Q_1' &= P_1 + \frac{1}{2}\frac{\partial}{\partial Q_2}(f^2 + g^2), \\[2mm]
Q_2' &= P_2 - \frac{1}{2}\frac{\partial}{\partial Q_1}(f^2 + g^2), \\[2mm]
Q_3' &= D.
\end{aligned}$$

The last equation is $dt = Dd\bar{t}$ as expected. Introducing

$$\phi(Q_1, Q_2) = f(Q_1, Q_2) + ig(Q_1, Q_2),$$

and

$$|\phi|^2 = f^2 + g^2,$$

the second set of the Hamiltonian equations of motion becomes

$$(40) \quad
\begin{aligned}
P_1' &= -P_3\frac{\partial D}{\partial Q_1} - \frac{1}{2}\left[P_1\frac{\partial^2}{\partial Q_1\partial Q_2} - P_2\frac{\partial^2}{\partial Q_1^2}\right]|\phi|^2 + \frac{\partial}{\partial Q_1}(D\widetilde{F}), \\[2mm]
P_2' &= -P_3\frac{\partial D}{\partial Q_2} - \frac{1}{2}\left[P_1\frac{\partial^2}{\partial Q_2^2} - P_2\frac{\partial^2}{\partial Q_2\partial Q_1}\right]|\phi|^2 + \frac{\partial}{\partial Q_2}(D\widetilde{F}), \\[2mm]
P_3' &= 0.
\end{aligned}$$

The last equation expresses the fact that $\widetilde{H} = $ constant. The singularity problem appears in the last term of the first two equations of the system (40) and it can be solved by selecting the proper D function, i.e., specifying f or g.

Equations (39) and (40) represent a sixth-order system, or omitting the last equation in both groups, we have a fourth-order system of equations which is equivalent with Equations (14). To show this the reader will have to compute Q_1'' and Q_2'' from Equations (39) and substitute in the resulting equations the values of P_1' and P_2' as given by Equations (40). The two second-order equations obtained this way are

(41)
$$Q_1'' - 2DQ_2' = \frac{\partial}{\partial Q_1}D\left(\frac{1}{2}|\phi|^2 - P_3 + \widetilde{F}\right),$$

$$Q_2'' + 2DQ_1' = \frac{\partial}{\partial Q_2}D\left(\frac{1}{2}|\phi|^2 - P_3 + \widetilde{F}\right).$$

For further comparison we recall that $|\phi|^2 = q_1^2 + q_2^2 = r^2$ by Equations (33). From this it follows that

$$\frac{1}{2}|\phi|^2 - P_3 + \widetilde{F} = \Omega - \frac{1}{2}\mu(1-\mu) + \widetilde{H},$$

or

(42)
$$\frac{1}{2}|\phi|^2 - P_3 + \widetilde{F} = \Omega - \frac{C}{2}.$$

Substituting Equation (42) into (41) gives

(43)
$$Q_1'' - 2DQ_2' = \frac{\partial}{\partial Q_1}D\left(\Omega - \frac{C}{2}\right),$$

$$Q_2'' + 2DQ_1' = \frac{\partial}{\partial Q_2}D\left(\Omega - \frac{C}{2}\right).$$

Finally, using u for Q_1, v for Q_2, $\lambda(Q_1, Q_2)$ for D, and Ω^* for $D(\Omega - C/2)$, we have

(44)
$$u'' - 2\lambda v' = \Omega_u^*,$$

$$v'' + 2\lambda u' = \Omega_v^*,$$

a special case of which is Equation (14), viz. $\lambda \equiv 1$.

Generalization of the problem. Equations (44) represent the restricted problem in the $(u, v, \bar{t})$ system which is obtainable from the (x, y, t) set of variables by the relations given in the previous chapter. It is essential to note that the type of regularizing transformations introduced does not change the general form of the equations. This remarkable invariance suggests the importance of the equations of the type given by (44). But one might go even further in the generalization of the problem. The class of dynamical problems described by the Lagrangian

$$(45) \qquad L = \frac{1}{2}(\dot{q}_1^2 + \dot{q}_2^2) + \dot{q}_1\alpha(q_1, q_2) + \dot{q}_2\beta(q_1, q_2) + \gamma(q_1, q_2)$$

leads also to equations like (44), since the equations of motion will contain only certain combinations of the three functions α, β, γ; i.e., the dynamics of the system can be represented by two functions only—corresponding to λ and Ω^*.

Another example leading to Equations (44) is the elliptic restricted problem, also called the semirestricted problem. Here the primaries move on elliptic orbits instead of circles. By selecting proper variables this dynamical problem is also reducible to equations identical in form to Equations (44), with the important difference that the dependent variable appears explicitly in Ω^*.

Generalization of the variables which describe the dynamical system and generalization of the dynamical system itself leaves Equations (44) invariant, as I have demonstrated with some examples above. Equations (44) describe the conventional problème restreint in a Cartesian rectangular synodic coordinate system with $\lambda \equiv 1$, with appropriate Ω and with the time as the independent variable. If $\lambda = \lambda(u, v)$ and the time is transformed as described in the previous chapter, we obtain the regularized equations of motion. With $\lambda \equiv 1$ and with the true anomaly as the independent variable, the elliptic problem is represented. Finally with $\lambda = \frac{1}{2}(\beta_{q_1} - \alpha_{q_2})$ Equations (44) represent all dynamical systems whose Lagrangian is given by Equation (45).

The general importance of Equations (44) in dynamics was recognized by Poincaré and by G. D. Birkhoff. The presence of the first-derivative terms (referring to either Equations (44) or (14)) render these equations irreversible since changing the sign of the independent variable does not preserve the equations. The structure of the equations is such that as long as the independent variable does not appear explicitly in Ω, an integral of the system is always available for any λ function. This is not surprising since the appearance of the λ function is not an essential part of the dynamical description of the system. The major difficulty of the dynamical system originates partly from the fact that $\lambda \neq 0$ and partly from the complexity of the Ω (or Ω^*) function.

If the problem is changed into the discussion of a single-degree-of-freedom dynamical system, the situation is altered completely and one gets $\ddot{x} = f(x)$, an unquestionably uninteresting equation.

The essential features of two-degrees-of-freedom irreversible dynamical systems are therefore contained in the equations

$$\ddot{x} - 2\dot{y} = \Omega_x,$$

$$\ddot{y} + 2\dot{x} = \Omega_y.$$

The fact that these equations also describe the restricted problem of three bodies is rather significant since the restricted problem occupies a critical place in celestial mechanics being the simplest nonintegrable problem of definite physical importance.

References

1. G. D. Birkhoff, *Sur le problème restreint des trois corps,* Two memoirs, Ann. Scuola Norm. Sup. Pisa (2) 4(1935), 267-306; (2) 5(1936), 1-42.

2. _____ , *Dynamical systems with two degrees of freedom,* Trans. Amer. Math. Soc. 18(1917), 1-70.

3. _____ , *The restricted problem of three bodies,* Rend. Circ. Mat. Palermo 39(1915), 1-70.

4. K. Bohlin, Acta. Math. 10(1887), 115.

5. D. Brouwer and G. Clemence, *Methods of celestial mechanics,* Academic Press, New York, 1961, pp. 253-268.

6. C. Burrau, Vjs. Astronom. Ges. 41(1906), 261.

7. C. L. Charlier, *Die Mechanik des Himmels,* Leipzig, von Veit & Co., Vol. I, 1902, pp. 89-171; Vol. II, 1907, pp. 289-296.

8. V. Dainelli, Giorn. Math. 18(1880), 271.

9. G. H. Darwin, Acta Math. 21(1897), 99-242.

10. L. Euler, Mem. de Berlin 228(1960).

11. _____ , Theoria motuum lunae, Academiae Imperialis Scientiarum, Petropoli, 1772.

12. G. W. Hill, *Researches in the lunar theory,* Amer. J. Math. 1(1878), 5-26, 129-147, 245-260.

13. S. S. Huang, Astronom. J. 67(1962), 304.

14. C. G. J. Jacobi, Computes rendus 3(1836), 59.

15. J. Kevorkian, *Uniformly valid asymptotic representation for all times of the motion of a satellite in the vicinity of the smaller body in the restricted three-body problem,* Astronom. J. 67(1962), 204-211.

16. A. Klose, *Topologische Dynamik der interplanetaren Massen,* Vjs. Astronom. Ges. 67(1932), 61-102.

17. J. L. Lagrange, Oeuvres 6(1772), 229.

18. T. Levi-Civita, Acta Math. 42(1919), 99-144.

19. _____ , Acta Math. 30(1906), 305-327.

20. _____ , Ann. Math. (3) 9(1904), 1-32.

21. _____ , Ann. Math. (3) 5(1901), 221-309.

22. E. O. Lovett, Quart. J. Math. 42(1911), 252.

168

23. G. A. Merman, Trans. Inst. Theo. Astronom. Acad. USSR **8**(1961), 5.

24. P. J. Message, *Some periodic orbits in the restricted problem of three bodies and their stabilities*, Astronom. J. **64**(1959), 226-236.

25. F. R. Moulton, *Celestial mechanics*, Macmillan, New York, 1914, pp. 278-308.

26. ——, Proc. Math. Congr. Cambridge, England, Vol. II, pp. 182-187; also, *Periodic orbits*, Carnegie Inst., Washington, 1920.

27. H. Poincaré, *Les méthodes nouvelles de la mécanique céleste*, Gauthier-Villars, Paris, 1892, 1893, 1899.

28. E. Rabe, *Determination and survey of periodic Trojan orbits in the restricted problem of three bodies*, Astronom. J. **66**(1961), 500.

29. ——, Astronom. J. **67**(1962), 382.

30. E. Rabe and A. Schanzle, Astronom. J. **67**(1962), 732.

31. E. Strömgren, Bull. Astronom. (2) **9**(1935), 87-130; Publ. Coph. Obs. # 100(1935), 1-44.

32. C. F. Sundman, Acta Math. **36**(1912), 105-192.

33. V. Szebehely, Proc. 9th Congr. Int. Astronom. Fed. 431, 1960.

34. ——, Proc. 2nd Int. Space Sci. Sympos. 319, 1961.

35. ——, Astronautics **7**(1962), 52.

36. ——, *Zero velocity curves and orbits in the restricted problem of three bodies*, Astronom. J. (3) **68**(1963), 147-151.

37. ——, *On isotach orbits*, Arch. Rational Mech. Anal. **13**(1963), 192-205.

38. ——, *Generation of orbits by generalized Hill curves*, J. Franklin Inst. **275**(1963), 371-380.

39. ——, *Application of the restricted problem of three bodies to space mechanics*, Space Science Rev. (2) **2**(1963), 219-249.

40. V. Szebehely and G. Giacaglia, Astronom. J. **69**(1964), 230-235.

41. V. Szebehely, *Perturbations of the regularized equations of the restricted problem of three bodies*, Astronom. J. **69**(1964), 309-315.

42. T. N. Thiele, Astronom. Nachr. **138**(1895), 1-10.

43. B. Thüring, Astronom. Nachr. **280**(1951), 226.

44. E. T. Whittaker, *Analytical dynamics*, 4th ed., Dover, New York, 1944, pp. 353-356; 406-412.

45. A. Wintner, *The analytical foundations of celestial mechanics*, Princeton Univ. Press, Princeton, N. J., 1941.

Yale University Observatory
New Haven, Connecticut

George Contopoulos

Problems of
Stellar Dynamics

I. **Introduction.** Stellar dynamics is a relatively recent branch of dynamical astronomy. It has been developed mainly in the present century as a part of astrophysics. Many authors have considered dynamical problems in their study of stellar clusters, the Galaxy and the other galaxies, or clusters of galaxies.

However there are some books devoted exclusively to stellar dynamics as a special discipline.

Smart's book, *Stellar dynamics* [1], deals mainly with the kinematical aspects of stellar dynamics, namely the solar motion, the two star-streams and the ellipsoidal theory of stellar velocities, the statistics of the distribution of stars around the sun, etc. Only a minor part is devoted to the dynamical problems of stellar clusters and of the Galaxy. However this book includes a lot of useful material.

A classical book in stellar dynamics is Chandrasekhar's *Principles of stellar dynamics* [2]. Chandrasekhar's book is mainly devoted to his own work on stellar systems. He finds first a complete formula for the time of relaxation of a stellar system, and then he discusses at length the quadratic solutions of Liouville's equation. Some applications of his work on the dynamics of the clusters and the

Galaxy are very interesting. The Dover edition of this book contains also Chandrasekhar's discussion of dynamical friction and an exposition of his views concerning the statistical mechanics of stellar systems.

A broadly similar pattern is used by von der Pahlen in his *Einführung in die Dynamik von Sternsystemen* [3]. However, von der Pahlen insists more on the observational part of stellar dynamics.

A rather different approach is due to Kurth, *An introduction to the dynamics of stellar systems* [4]. Kurth discusses the three main approaches to the problems of stellar dynamics, the n-body problem approach, the continuum approach, and the statistical approach. He gives a number of interesting theorems concerning the first two approaches (some of them not well known before), but he disregards completely the third approach, which he considers as impossible.

Galactic dynamics has been considered by B. Lindblad in a very good review paper in the Handbuch der Physik [5]. A rather short review, *Recent developments in stellar dynamics*, is given by Camm in Vistas in Astronomy [6].

Two Russian books on this subject have appeared lately. One is Ogorodnikov's *The dynamics of stellar systems* [7]; this is to be published in English by Pergamon Press. It deals with the kinematics and dynamics of stellar systems in general and of the Galaxy in particular.

The second book is Idlis' *Structure and dynamics of stellar systems* [8]. It deals with the dynamics of the Galaxy and of the other galaxies, considered as stationary stellar systems.

In general all these authors consider stellar dynamics quite separately from classical celestial mechanics. However, both stellar dynamics and celestial mechanics are parts of dynamical astronomy. Their subject matter is essentially the same—the gravitational interaction of a number of material bodies—and some of their methods are similar.

Of course there is one important difference between them. The number of gravitating bodies in the case of stellar dynamics is much greater than in celestial mechanics, and this introduces a number of new and very difficult problems. In order to discuss these new problems, new methods have to be introduced in many cases. However, there are many cases where stellar dynamics can profit from the methods and experience of the classical celestial mechanics and, likewise, methods developed in stellar dynamics can be applied to problems of the solar system.

This interchange of ideas has not been done to a satisfactory degree until now. It would be profitable to bring these two disciplines much more closely together into a unified whole.

The present chapter contains essentially my lectures at the Yale University Observatory in 1962. It deals with some of the major problems in this field and indicates the points where research is under way or should be desirable. Further it gives the necessary references until 1962, or 1963 in some cases.

Three different techniques have been used in most work in stellar dynamics. Our treatment will discuss these more or less separately, and we now indicate what the three are.

(a) *The n-body problem approach.* Any stellar system is composed of n gravitating bodies, interacting according to the law of Newton. The equations of motion can be solved numerically if the number of stars is small, and this often gives an adequate solution of the problem. Such integrations have been done lately by S. von Hörner [9], [10] and have given interesting results.

However, if the number of the stars is great, even the most powerful electronic computers are unable to calculate their motions. On the other hand one can find a number of general theorems about a system of n bodies that can give us some information about the behavior of the system in general. Such theorems are the virial theorem and the theorems of Poincaré and Hopf. Unfortunately no complete description of the behavior and the evolution of stellar systems can be effected by the available theorems. Therefore the most accurate method of dealing with the problems of stellar dynamics is, in general, insufficient.

(b) *The continuum approach.* When the number of stars in a stellar system is big enough we may consider the system as a gravitating continuum, i.e., we may consider the smoothed out gravitational field that we would have if the stars were pulverized and their matter distributed evenly in the interstellar space. More accurately, we consider that the phase space is evenly filled by points representing the positions and velocities of the stars in the actual space.

This procedure is justified as regards the action of the distant stars on any single star of a stellar system; because in that case the potential of a great number of stars is almost the same as that of a gravitating continuum. However, this approach neglects the interaction of a star with its neighbors. In a close approach of two stars, their motion is influenced mainly by each other and the action of the distant stars may be considered only as a perturbation. This is the main limitation of the continuum approach.

However, in most stellar systems the time of relaxation is very great, so that in the mean the close encounters have a relatively small effect on the dynamics of the stars. In general a star can make at least some revolutions (sometimes a great number of revolutions) around the center of the system before its orbit is changed appreciably because of close encounters with other stars.

(c) *The statistical approach.* In most cases statistical methods are used in stellar dynamics. The calculation of the time of relaxation or of the mean free path of a stellar system, and the study of the distribution of velocities or of the density distribution of a cluster, require such methods. It is true that this approach has one basic difficulty; a consistent definition of probability is, in most cases, not given. However the statistical methods have given many interesting results about stellar systems in general. The verification of these results is one of the main problems of stellar dynamics at the present moment.

II. The n-body problem approach.

1. *The Virial Theorem.* The classical virial theorem is due to Lagrange and Jacobi. Lagrange was the first to apply it in the problem of three bodies; Jacobi applied it to the n-body problem.

The equations of motion of the n-body problem are

$$m_i \ddot{x}_i = -\frac{\partial V}{\partial x_i},$$

(1) $$m_i \ddot{y}_i = -\frac{\partial V}{\partial y_i}, \qquad (i = 1, 2, \cdots, n),$$

$$m_i \ddot{z}_i = -\frac{\partial V}{\partial z_i},$$

where

(2)
$$V = -\frac{G}{2} \sum_{i=1}^{n} \sum_{j=1}^{n} \frac{m_i m_j}{r_{ij}} \qquad (i \neq j)$$

is the total potential energy of the system and r_{ij} is the distance between the stars i and j.

Let the distances of the n bodies from their center of mass be r_i; then the moment of inertia about the center of mass is

(3)
$$J = \sum_{i=1}^{n} m_i \bar{r}_i^2.$$

By differentiating it twice we find

(4)
$$\frac{1}{2} \ddot{J} = \sum_{i=1}^{n} m_i \dot{\bar{r}}_i^2 + \sum_{i=1}^{n} m_i \bar{r}_i \ddot{\bar{r}}_i.$$

But

(5)
$$\sum_{i=1}^{n} m_i \dot{\bar{r}}_i^2 = 2T,$$

where T is the kinetic energy of the system with respect to its center of mass.

Further

(6)
$$\sum_{i=1}^{n} m_i \bar{r}_i \ddot{\bar{r}}_i = \sum_{i=1}^{n} m_i (x_i \ddot{x}_i + y_i \ddot{y}_i + z_i \ddot{z}_i)$$
$$= -\sum_{i=1}^{n} \left(\frac{\partial V}{\partial x_i} x_i + \frac{\partial V}{\partial y_i} y_i + \frac{\partial V}{\partial z_i} z_i \right)$$

and this last quantity is equal to V, because V is homogeneous of degree -1 with respect to x_i, y_i, z_i.

Hence

(7)
$$\frac{1}{2} \ddot{J} = 2T + V,$$

and as $T + V = E$ (the total energy), we get

(8)
$$\frac{1}{2} \ddot{J} = E + T = 2E - V.$$

This is the Lagrange equality.

If a system is stationary, so that its form does not change in the mean, then $\ddot{J} = 0$ and

$$(9) \qquad\qquad 2T + V = 0.$$

This equation is usually called the Virial Theorem. A more general form of the same equation is

$$(10) \qquad\qquad \langle 2T \rangle + \langle V \rangle = 0,$$

where the symbol $\langle \ \rangle$ means time average for $t \to \infty$. This equation is valid when a system remains bounded. It would be valid also in a system expanding slowly, so that $\dot{J} \to$ const. ($\ddot{J} \to 0$). But such a case is probably exceptional.

Some important consequences can be drawn from the Lagrange equation. If $E > 0$, then $\frac{1}{2}\ddot{J} > E > 0$.

Hence $\dot{J} > 2Et + c$ and $J > Et^2 + ct + c'$, where c and c' are constants. Therefore $J \to \infty$ as $t \to \infty$, and if $\overline{m}$ is the greatest mass of any star,

$$(11) \qquad\qquad \overline{m} \sum_{i=1}^{n} r_i^2 \geqq J;$$

i.e., the distance of the representative point (x_i, y_i, z_i) (in the space of $3n$ dimensions) from the origin tends to infinity.

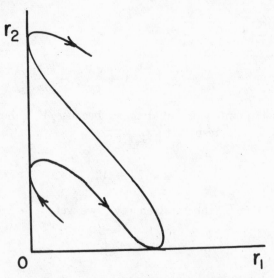

FIGURE 1. Oscillations of Two Stars

This does not necessarily mean that one star will actually escape from the system; in fact, although the sum of the squares of the distances of all the stars from the center of mass increases continuously, the individual distances may oscillate in such a way that no star actually escapes from the system. This fact is illustrated by Figure 1, where the distances of two stars from the center of mass oscillate alternatively with increasing oscillations. In real systems once a star has gone very far, it comes under the influence of other fields and can be considered as lost from the original system. However it would be of great theoretical interest to determine under what conditions oscillations of the form of Figure 1 are possible in an isolated cluster.

It has been assumed by some authors that the escape of stars from a cluster must be effected in pairs or even triples, in order that the center of mass should not move. However this is not necessary. It is most probable that only one star is ejected at a time, while the rest of the cluster recedes with respect to the center of mass so that the equalities

$$\sum_{i=1}^{n} m_i x_i = \sum_{i=1}^{n} m_i y_i = \sum_{i=1}^{n} m_i z_i = 0$$

persist.

If $E > 0$, and one star is ejected, it is most probable that the energy of the remaining cluster with respect to its new center of mass is again positive; therefore another star also will probably be ejected, and so on. It is for this reason that stellar systems of positive energy are disintegrating. Such systems are the associations which are expanding with rather big velocities and disintegrate completely in some millions of years.

If $E < 0$ there is almost no known criterion to decide whether a stellar system will eject stars or not. A partial answer to this problem is given by Hopf's theorem, that is mentioned below. Further work on this problem has been done by Chazy [11], [12], [13] in the case of the three-body problem, and by the Russian authors Khilmi, Merman, Schmidt, etc. A book by Khilmi, *Qualitative methods in the many body problem*, has recently been translated into English and German [14]. One can find there references to the Russian literature on the n-body problem. Further a *Review of*

Soviet celestial mechanics literature including a great amount of material has been published by the U.S. Department of Commerce [15].

A number of systematic generalizations of the classical virial theorem have been given lately by Chandrasekhar [16], [17]. These are the tensor virial theorem, and the virial theorems of order higher than the second. The virial theorem for the n-bodies in the post-Newtonian approximation of general relativity has been considered by Chandrasekhar and Contopoulos [18].

If we multiply the first equation of the system (1) by y_i and add all the terms with different i we get:

$$(12) \qquad \sum_{i=1}^{n} m_i \ddot{x}_i y_i = - \sum_{i=1}^{n} y_i \frac{\partial V}{\partial x_i}$$

or

$$\sum_{i=1}^{n} m_i \frac{d}{dt} (\dot{x}_i y_i) - \sum_{i=1}^{n} m_i \dot{x}_i \dot{y}_i = - \sum_{i=1}^{n} m_i y_i \sum_{j=1; j \neq i}^{n} \frac{Gm_j(x_i - x_j)}{r_{ij}^3}$$

$$(13)$$

$$= - \frac{1}{2} \sum_{i=1}^{n} \sum_{j=1; j \neq i}^{n} \frac{Gm_i m_j(x_i - x_j)(y_i - y_j)}{r_{ij}^3}.$$

If we write

$$(14) \qquad - \frac{1}{2} \sum_{i=1}^{n} \sum_{j=1; j \neq i}^{n} \frac{Gm_i m_j(x_i - x_j)(y_i - y_j)}{r_{ij}^3} = V_{xy}$$

and

$$(15) \qquad \frac{1}{2} \sum_{i=1}^{n} m_i \dot{x}_i \dot{y}_i = T_{xy},$$

then

$$(16) \qquad 2 T_{xy} + V_{xy} = \sum_{i=1}^{n} m_i \frac{d}{dt} (\dot{x}_i y_i).$$

V_{xy} and T_{xy} are the potential energy tensor and the kinetic energy tensor respectively.

In general if we multiply the first equation of the system (1) by $x_i^a y_i^b z_i^c$ and add the corresponding terms we find

$$(17) \qquad \sum_{i=1}^{n} m_i \ddot{x}_i x_i^a y_i^b z_i^c = - \sum_{i=1}^{n} x_i^a y_i^b z_i^c \frac{\partial V}{\partial x_i},$$

or

$$\frac{d}{dt} \sum_{i=1}^{n} m_i \dot{x}_i x_i^a y_i^b z_i^c - a \sum_{i=1}^{n} m_i \dot{x}_i^2 x_i^{a-1} y_i^b z_i^c$$

$$- b \sum_{i=1}^{n} m_i \dot{x}_i \dot{y}_i x_i^a y_i^{b-1} z_i^c - c \sum_{i=1}^{n} m_i \dot{x}_i \dot{z}_i x_i^a y_i^b z_i^{c-1}$$

$$(18)$$

$$= - \sum_{i=1}^{n} \sum_{j=1; j \neq i}^{n} \frac{G m_i m_j x_i^a y_i^b z_i^c (x_i - x_j)}{r_{ij}^3}$$

$$= - \frac{1}{2} \sum_{i=1}^{n} \sum_{j=1; j \neq i}^{n} \frac{G m_i m_j (x_i^a y_i^b z_i^c - x_j^a y_j^b z_j^c)(x_i - x_j)}{r_{ij}^3}.$$

But if we set

$$(19) \quad x_i = x_j + (x_i - x_j), \quad y_i = y_j + (y_i - y_j), \quad z_i = z_j + (z_i - z_j),$$

we find

$$x_i^a y_i^b z_i^c - x_j^a y_j^b z_j^c$$

$$(20)$$

$$= \sum_{p=1}^{a-1} \sum_{q=1}^{b-1} \sum_{r=1}^{c-1} C_p^a C_q^b C_r^c x_i^{a-p} y_i^{b-q} z_i^{c-r} (x_i - x_j)^p (y_i - y_j)^q (z_i - z_j)^r.$$

Thus if we write

$$(21) \qquad \sum_{i=1}^{n} m_i \dot{x}_i^a \dot{y}_i^b \dot{z}_i^c x_i^d y_i^e z_i^f = T_{abc;def}$$

and

$$(22) \quad - \frac{1}{2} \sum_{i=1}^{n} \sum_{j=1; j \neq i}^{n} \frac{G m_i m_j (x_i - x_j)^a (y_i - y_j)^b (z_i - z_j)^c x_i^d y_i^e z_i^f}{r_{ij}^3} = V_{abc;def},$$

we find

$$\frac{d}{dt} T_{100;abc} = a T_{200;a-1,bc} + b T_{110;a,b-1,c} + c T_{101;ab,c-1}$$

$$(23)$$

$$+ \sum_{p=1}^{a-1} \sum_{q=1}^{b-1} \sum_{r=1}^{c-1} C_p^a C_q^b C_r^c V_{p+1,q,r;a-b,b-q,c-r}.$$

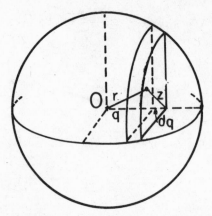

FIGURE 2. Integration Element in a Star Cluster

(a) *Applications of the Virial Theorem*. A well-known application of the classical virial theorem is for the determination of the mass of a cluster in a stationary state. If we define a mean square velocity by the formula $T = M\overline{v^2}/2$ where M is the total mass, Equation (9) for a system of stars of equal masses gives

$$(24) \qquad M\overline{v^2} = \frac{G}{2} \sum_{i=1}^{n} \sum_{j=1;\, j \neq i}^{n} \frac{m_i m_j}{r_{ij}} = G \frac{n(n-1)\, m^2}{r^*} \simeq \frac{GM^2}{r^*},$$

where r^* is defined by

$$(25) \qquad \frac{n(n-1)}{r^*} = \frac{1}{2} \sum_{i=1}^{n} \sum_{j=1;\, j \neq i}^{n} \frac{1}{r_{ij}}.$$

We call r^* the radius of the cluster; this is defined accurately enough in the case of spherical clusters by a method devised by M. Schwarzschild [19] which is based on star counts on a photograph of the cluster([1]).

Let the number of star images between two parallel straight lines whose distances from the center are q and $q + dq$ be $S(q)\, dq$ (Figure 2). These images are the projections of all the stars between two parallel planes. On each plane the curves of constant density ρ are circles of radius $z = (r^2 - q^2)^{1/2}$. Thus

---

([1]) Chandrasekhar, Kurth and others give a radius R^* that is approximately equal to one half of r^*.

(26)
$$S(q)\,dq = \int_{z=0}^{\infty} \rho\, 2\pi z \, dz \, dq.$$

But we have $zdz = rdr$; therefore

(27)
$$S(q) = \int_{q}^{\infty} \rho\, 2\pi r dr.$$

Instead of ∞ we may set as upper limit a visible boundary of the cluster, say r_{Max}.

The gravitational potential of the cluster is

(28)
$$V = -\int_{0}^{\infty} \frac{GM(r)}{r}\, 4\pi r^2 \rho\, dr = 2G \int_{0}^{\infty} M(q)\, dS(q),$$

because

(29)
$$dS(q) = -\rho(q)\, 2\pi q \, dq.$$

Hence

(30)
$$V = [2\, GM(q)\, S(q)]_0^{\infty} - 2G \int_{0}^{\infty} S(q)\, 4\pi q^2 \rho dq.$$

The first term is zero because $M(0) = 0$ and $S(\infty) = 0$ (at the boundary). Therefore

$$V = -4G \int_{0}^{\infty} q S(q)\, dS(q)$$

(31)
$$= [2GqS^2(q)]_0^{\infty} - 2G \int_{0}^{\infty} S^2(q)\, dq$$

$$= -2G \int_{0}^{\infty} S^2(q)\, dq.$$

On the other hand

$$M = \int_{0}^{\infty} 4\pi r^2 \rho\, dr = -2 \int_{0}^{\infty} q\, dS(q)$$

(32)
$$= [-2q S(q)]_0^{\infty} + 2 \int_{0}^{\infty} S(q)\, dq$$

$$= 2 \int_{0}^{\infty} S(q)\, dq.$$

Hence the formula

(33) $$V = -\frac{GM^2}{r^*}$$

gives

(34) $$r^* = \frac{2(\int_0^\infty S\,dq)^2}{\int_0^\infty S^2\,dq}.$$

From equation (24) we get

(35) $$M = \frac{r^*\overline{v^2}}{G} = \frac{3r^*\overline{v_r^2}}{G},$$

where $(\overline{v_r^2})^{1/2}$ is the root mean square radial velocity, that is measured spectroscopically. This method has been used successfully for finding the masses of the globular clusters.

The same method has been used lately to find the masses of clusters of galaxies. There, however, a serious discrepancy has been observed. If M is the mass of the cluster and n the number of the galaxies, the mass of each galaxy is calculated from the formula $m = M/n$. On the other hand the masses of the individual galaxies can be found by measuring spectroscopically the rotational velocity of each galaxy with respect to its own center of mass. However the masses found by the latter method are much smaller than the masses calculated by means of the virial theorem. Two main explanations of this discrepancy have been proposed. One is that the virial theorem is not applicable in this case because the system is rapidly expanding, so that $\ddot{J} > 0$ (not only $\dot{J} > 0$). Then

$$\frac{1}{2}\ddot{J} = M\left(\overline{v}^2 - \frac{GM}{r^*}\right) > 0 \text{ and } M < \frac{r^*\overline{v^2}}{G}.$$

This view has been advocated by Ambartsumian [20]. The second explanation is that the clusters contain a great amount of intergalactic matter (or the galaxies extend much further than the limits that can be reached spectroscopically); therefore the mass of the cluster is bigger than the mass of the observable parts of its galaxies, $M > nm$.

A lengthy discussion of these views took place at the Santa Barbara Conference on the Instability of Systems of Galaxies in 1961 [21]. No definite conclusion has been reached yet.

Chandrasekhar and Lebovitz [22], [23] have applied the tensor virial theorems in problems concerning the equilibrium of rotating fluids.

A few applications of generalized virial theorems to nonspherical systems have been given by van Wijk [24] and King [25]. Further extension of this work should be useful.

Limber [26], [27], [28] has given a form of virial theorem that takes account of the interstellar matter also.

2. *Poincaré's Recurrence Theorem*. One of the most important theorems concerning the long range evolution of certain dynamical systems is the "recurrence theorem" of Poincaré. This theorem has been given by Poincaré in his *Méthodes nouvelles de la mécanique céleste* [29, Chapter 26]. Its most accurate form, however, is due to Carathéodory [30].

Suppose that we have a continuous, measure-preserving flow in a space of N dimensions, defined by the equations

$$x_i = \phi_i(x_0, t),$$

where $x_0(x_{10}, x_{20}, \cdots, x_{N0})$ gives the initial conditions for $t = 0$. We assume that $\phi(\phi(x_0, t_1), t_2) = \phi(x_0, t_1 + t_2)$, where ϕ represents the set of functions $\phi_1, \phi_2, \cdots, \phi_N$.

Also let there be an invariant set Ω with finite measure M (i.e., the trajectories issuing from all the points of Ω are contained inside Ω). We can assimilate Ω to a closed vessel inside which moves an incompressible fluid.

We shall prove now that the trajectories issuing from almost any point x_0 of Ω come an infinite number of times in the neighborhood of x_0. More accurately, x_0 is a point of accumulation of the points (images) $\phi(x_0, \tau)$, $\phi(x_0, 2\tau) \cdots$, where τ is a definite positive or negative number. "Almost any" means that the measure of the exceptional points, where this property does not hold, is zero.

Suppose that we have a set w of points x_0 of measure m, inside Ω. The set of the points $\phi(x_0, \tau)$, where $x_0 \in w$, has the same measure m. Let this set be written $\phi(w, \tau)$. We shall prove that any set w of measure $m > 0$, has points in common with $\phi(w, q\tau)$, for some value of q.

In fact, if w and $\phi(w, \tau), \phi(w, 2\tau), \cdots, \phi(w, q\tau)$ have no points in common, then their total measure is $(q + 1) m$. This, however, is greater than M if $q + 1 > M/m$; therefore some sets of the above sequence have a common part, e.g., $\phi(w, q_1\tau)$ and $\phi(w, q_2\tau)$. But we know that $\phi(\phi(x_0, (q_2 - q_1)\tau), q_1\tau) = \phi(x_0, q_2\tau)$, i.e., to the above common part there corresponds a common part between w and $\phi(w, (q_2 - q_1)\tau)$. (See Figure 3.)

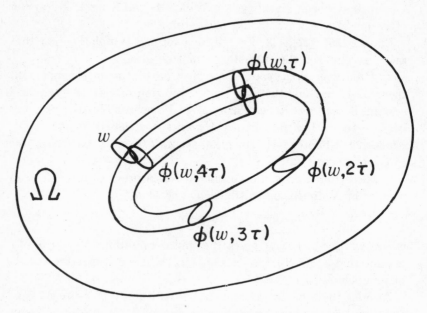

FIGURE 3. Overlapping Sets

Suppose now that A_d is the set of the points x_0 all of whose images have a distance from x_0 greater than $d > 0$. We shall prove that the measure of A_d is zero. In fact, if $m(A_d) > 0$ we may separate A into a number of sets whose greatest dimension is smaller than d. Then the measure of at least one of these sets, w, is greater than zero. But we have proved that this set has points in common with $\phi(w, q\tau)$ for some q. This means that for some points of w the distance from their images is smaller than d, contrary to our assumption. Hence $m(A_d) = 0$. This is true for each d; therefore the measure of the points whose images do not come arbitrarily near to them is zero.

The theorem of Poincaré does not mean that such exceptions do not occur. It means only that the probability of these exceptions is zero, with the measure-theoretical definition of probability.

3. *The Theorems of Hopf*. If there is no invariant set Ω of finite measure then we have a generalization of Poincaré's theorem due to Hopf [31].

Hopf calls "escape points" the points whose images $\phi(x_0, \tau)$, $\phi(x_0, 2\tau), \cdots$ have no accumulation point in Ω. Then according to the Weierstrass-Bolzano theorem these images will have a point of accumulation at the boundary of Ω. On the other hand the points that are accumulation points of their images are called "recurrent points." Then Hopf's first theorem states that almost all the points of Ω are either recurrent or escape points. It can be proved further that the escape points which do not go to infinity form a set of measure zero. Thus almost all points of Ω are either recurrent, or they go to infinity.

Of more importance is Hopf's second theorem. In order to state this theorem we distinguish between the two directions of time and call a point recurrent or escape point with respect to the past or the future. The second theorem of Hopf states that almost all the points that are escape (or recurrent) points with respect to the past are also escape (or recurrent) points with respect to the future.

Let B be the set of the escape points with respect to the past inside any bounded set $U_i \subset \Omega$. B is the union of the sets B_n, where B_n is the set of points x_0, whose images $(x_0, -q\tau)$ are all outside U_i for $q > n$. Then B_n has no points in common with the sets $\phi(B_n, -q\tau), \phi(B_n, -2q\tau), \cdots$, and consequently any two such sets have no point in common. It follows that B_n, $\phi(B_n, q\tau), \phi(B_n, 2q\tau), \cdots$ have no point in common either; for otherwise B_n should have points in common with some set $\phi(B_n, -q'q\tau)$. Then it can be proved that almost all points of B_n are escape points with respect to the future; in fact if we separate Ω into bounded sets U_i, there cannot be a subset of points of B_n of positive measure that have accumulation points in any U_i, because then some images of B_n should overlap, according to the first part of Poincaré's theorem and we have seen that this is impossible.

We conclude that almost all the escape points with respect to the past are also escape points with respect to the future. Further, as almost all points are either recurrent or escape (in both directions of time), almost all recurrent points with respect to the past are also recurrent points with respect to the future.

Hopf's theorem does not exclude orbits of the form 4a that are neither recurrent nor escape, or 4b, that are escape points with respect to the future only (see Figure 4), but states that such orbits have a probability zero.

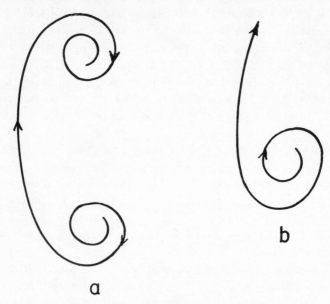

FIGURE 4. Exceptional Orbits

4. *Applications of Poincaré's and Hopf's Theorems.* Poincaré's theorem can be applied to the motion of particles in a stationary field only if there is an invariant set of finite measure. In general this happens if the potential energy V has a lower bound. The set Ω of points whose total energy is between two finite limits $a < E < b$ is invariant, because the energy E is invariant along any trajectory of the representative point in phase space (i.e., the point that represents the positions and velocities of all the particles). If V has a lower bound V_B, then

$$V_B \leqq V = E - T < b - T \leqq b;$$

hence

$$T < - V_B + b,$$

i.e., $|x_i|$, $|y_i|$, $|r_i|$ are all bounded.

Further, $V_B \leqq V < b$, and if we take $b < V_\infty$, then the space coordinates that satisfy this inequality are bounded; hence Ω is bounded.

We must also state that in this case the flow in the phase space[2] Γ of the system is measure preserving. We have

(36) $$\frac{dx_i}{dt} = \dot{x}_i, \quad \frac{d\dot{x}_i}{dt} = -\frac{\partial V}{\partial x_i}, \qquad (i = 1, 2, \cdots, n)$$

and the expression

$$\sum_{i=1}^{n} \left\{ \frac{\partial \dot{x}_i}{\partial x_i} + \frac{\partial \dot{y}_i}{\partial y_i} + \frac{\partial \dot{z}_i}{\partial z_i} - \frac{\partial}{\partial \dot{x}_i}\left(\frac{\partial V}{\partial x_i}\right) - \frac{\partial}{\partial \dot{y}_i}\left(\frac{\partial V}{\partial y_i}\right) - \frac{\partial}{\partial \dot{z}_i}\left(\frac{\partial V}{\partial z_i}\right) \right\}$$

is zero, because all the terms are zero. This is a special case of the following theorem: If in a system of N differential equations $\dot{x} = f(x,t)$ we have the condition

$$\sum_{i=1}^{n} \frac{\partial f_i}{\partial x_i} = 0, \text{ then } \Delta = \frac{D(x_1, x_2, \cdots, x_N)}{D(x_{10}, x_{20}, \cdots, x_{N0})} = 1,$$

where x is a solution of these equations and x_0 the initial conditions. This theorem is easily proved if we remark that

$$\frac{\partial \Delta}{\partial t} = \Delta \sum_{i=1}^{n} \frac{\partial f_i}{\partial x_i} = 0.$$

Then the measure $\int_M dx$ (in N dimensions) of a set of points that occupy initially a set of measure $\int_{M_0} dx_0$ is $\int_M dx = \int_{M_0} \Delta \, dx_0$ $= \int_{M_0} dx_0$, i.e., the flow is measure preserving (incompressible).

In the case of the n-body problem we have a measure preserving flow; however, the potential V has no lower bound. In fact, near the gravitating bodies $V \to -\infty$, and the corresponding velocities may increase beyond any bound; therefore in the n-body problem there seems to be no invariant set of finite measure.

However there is at least one exception: this is the case of the plane restricted 3-body problem [32, p. 206-207]. Then $E = (\dot{x}_1^2 + \dot{x}_2^2)/2 + V$, where $V = (r^2/2) - (\mu/r_1) - ((1 - \mu)/r_2)$, $(r, r_1, r_2$ are the distances of the moving point from the origin and from the masses μ and $(1 - \mu)$ respectively). If r, r_1, r_2 are constant,

---

[2] We distinguish between the phase space of the molecule (μ-space, Boltzmann), that has $2r$ dimensions, where r are the degrees of freedom of each molecule, and the phase space of the system (Γ-space, Gibbs) that has $2rn$ dimensions.

V is constant. Then the inequalities $a < E < b$ are written $2(a - V) < \dot{x}_1^2 + \dot{x}_2^2 < 2(b - V)$; i.e., the point $(\dot{x}_1, \dot{x}_2)$ is inside a circular annulus of measure $\pi([2(b - V)] - [2(a - V)]) = 2\pi(b - a)$, and this is independent of x_1, x_2. Hence if the x_1, x_2 space is finite (this happens e.g., near each one of the two masses, inside the corresponding closed Hill's curves, where V is smaller than a certain value c), then the invariant set Ω in the phase space has a finite measure; therefore Poincaré's theorem can be applied.

However the corresponding measure of the $\dot{x}_1$, $\dot{x}_2$, $\dot{x}_3$ space in 3 dimensions tends to infinity as $V \to \infty$.

Some recent results of the Russian mathematician Arnol'd [33] indicate that in most cases of the 3 or n-body problem in 2 or 3 dimensions there are boundaries in finite distance that the moving point cannot cross. Therefore in these cases Poincaré's theorem is applicable and the system is recurrent.

In the most general case we can apply Hopf's first theorem and state that in the n-body problem either the system is recurrent so that it returns an infinite number of times near its initial state, or the representative point goes to infinity. The exceptional cases of nonrecurrent systems whose corresponding points do not go to infinity, have a probability zero. Hopf's second theorem now states that the corresponding points of almost every system will go to infinity if and only if they have come from infinity. This means that if a star escapes from a system, then either this star or another one was captured by the system, or the system was initially oscillating in the opposite way to that shown in Figure 1. Similarly the capture of a star by a cluster is, in general, temporary, and the system ejects later one star (not necessarily the captured one), or it becomes oscillating in the above sense.

A similar theorem was proved earlier by K. Schwarzschild [34] who found that the captures of comets by Jupiter are only temporary([3]).

Another consequence of Poincaré's and Hopf's theorems is that a stellar system cannot tend towards a special final form unless it

---

([3]) See also Chazy [13, p. 421ff.]. Chazy asserts that if a third body approaches a binary system from infinity it cannot form a stable triple system, or cause the disruption of the binary, or even be captured by one component of the binary while the other component escapes. However it seems that the last two possibilities exist. L. Becker [35], [36] calculated some orbits in the 3-body problem that resulted in the capture of a third body by a binary, while one of the components is either stationary or it is disintegrating. E.g., a rotating system cannot become flatter, if only conservative forces are applied. In fact, the flattening of the stellar systems must be ascribed to the friction of the prestellar gas. We mention some other applications of Poincaré's theorem in §III.3. These examples show the usefulness of topological methods in some problems of stellar dynamics.

5. *Some Problems.* Some unsolved problems should be mentioned in connection with the above theorems.

(a) If for very small distances the law of attraction of two stars is changed so that V does not tend to infinity, then one might apply Poincaré's theorem and state that all stellar systems are recurrent. But then the stars should not approach beyond a certain limit; e.g., they should never collide physically. This shows the connection between the problems of escape and of very close approaches in a stellar system. It should be of the greatest interest to find under what conditions the distances of the stars of a given system never become smaller than a given amount.

(b) As the criterion of Hopf cannot be verified in general (except in the case of recent captures), we need more practical criteria in order to determine whether a given system will eject stars or not.

By applying the statistical methods of the third chapter we shall see that in general a stellar system will eject stars. However, as the dynamical phenomena are time reversible, this means that it is equally likely to have an ejection in the inverse direction of time, i.e., a capture. This is the main idea underlying Hopf's second theorem. But it seems that the phenomenon of escape is very rare, and one should not disregard the possibility that in some cases of isolated clusters it never happens. In fact the numerical calculations of von Hörner, that we shall mention later, show that escapes of stars are much rarer than it was generally believed until now.

There are some reasons why the escape of stars is a very improbable phenomenon. If the velocity of a star gradually increases because of close encounters, the star moves to the outer parts of

---

of the binary escaped. Further O. Schmidt, Khilmi and other Russian authors proved, by means of numerical calculations also, that a third body can cause the disruption of a binary. If then we invert the time direction we find that 3 stars moving in special hyperbolic orbits can form a binary (see [14], [15], [37]). This phenomenon has a probability greater than zero. Similar results were found by Sibahara [38], [39], [40]. This work should be extended further.

the cluster where the encounters are much more rare, and therefore, as pointed out by Hénon [41], [42], it cannot really escape. Only in the case of a very close encounter can a star acquire abruptly the escape velocity.

If, however, a quasi-stable state is reached in a stellar system, it may be that such very close encounters never occur. The problem is worthy of further consideration.

(c) It should be of interest to find in what cases the corresponding points of stellar systems are recurrent points that go also to infinity. The existence of such oscillating orbits has·been proved lately by Sitnikov [43]. Sitnikov found a case of the 3-body problem where one body performs oscillations of greater and greater (unbounded) amplitudes, but it returns infinitely many times near the origin. Further, one should find under what circumstances two stars may oscillate alternatively, as in Figure 1.

Some very interesting work in classifying the different types of evolution of a system of three bodies was done by Chazy [11], [12], [13]. A number of Russian authors have also worked in this field (see [14], [15]). Further work of this type in the n-body problem should be very important.

(d) A mean recurrence time can be calculated for recurrent systems as follows: If M is the measure of an invariant set and m the measure of a sphere within which the point representing the system in the phase space must return, then the mean recurrence time is $M\tau/m$, where $\tau = d/v$ (d is the diameter of the sphere and v the "velocity" of the moving point).

Can such a mean recurrence time be defined in the case that the invariant measure M is infinite, and, if so, what is its order of magnitude?

Consider for example two clusters, an open cluster with 100 stars and a globular cluster with 10^5 stars, and let their respective diameters be 10pc and 100pc respectively. Their mean escape velocities are approximately 0.4 km/sec and 4 km/sec. Suppose that all the stars are always contained inside the above limits of the clusters and their velocities are smaller than the corresponding escape velocities. Then the recurrence time needed for all the stars to return within 2 A.U. from their initial positions with velocities differing less than 0.1 km/sec from their initial values is greater than 10^{1000} years in the first case and greater than $10^{1,000,000}$ years in the second case. Some consequences of this fact will be mentioned in the third chapter.

(e) It would be of the greatest importance to apply numerical methods to find some characteristics of a stellar system which would give indications concerning its evolution. Such methods were used by von Hörner [9], [10] and we shall mention them presently.

6. *Numerical Calculations of the n-body Problem.* Von Hörner [9], [10] calculated numerically the evolution of some clusters with a small number of stars, up to 25. Unfortunately it is very difficult to calculate the evolution of a cluster with a great number of stars because the time needed for such calculations is prohibitively great. In fact the time needed by von Hörner to calculate the evolution of a cluster for one time of relaxation is proportional to n^4. He spent 40 minutes (2/3 hours) for calculating the evolution of a cluster of 16 stars for one time of relaxation with the computer "Siemens 2002." Therefore for 160 stars one should need 270 days with the same method and the same computer.

However, even with a small number of stars, one can find many interesting results.

The initial positions and velocities used by von Hörner were a random distribution within a sphere in coordinate space and in velocity space correspondingly. Hence an initially constant density was assumed.

After some time, equal to 2 or 3 relaxation times, a somewhat stable distribution of density and velocity was established but this distribution changed gradually afterwards.

Although the first calculations (namely those in [9]) indicated a Maxwellian velocity distribution, it was found later (see [10]), when a greater number of stars was used, that this distribution was not even approximately Maxwellian. There were a greater number of large and small velocities than expected. The high-velocity stars were not escaping but formed close binaries near the center of the system. On the other hand the excess of small velocities corresponds to stars in the outer parts of the clusters.

The number of escaping stars was almost 8 times smaller than expected. In general the stars going to the outer parts of the system were fewer than expected. This fact indicates that escapes are much more rare than usually assumed.

In the first calculations the density of the central parts of the cluster was represented by an isothermal gas sphere. However later (see [10]) it was found that a central condensation is formed and the density follows there the law $1/r^2$. In the outer parts the density follows approximately the law $1/r^4$, as expected theoretically (see [44]).

Chandrasekhar's formula for the time of relaxation was verified. The virial theorem was also approximately satisfied, i.e., $V/2T \cong 1$ within a factor 2.

These results show that the statistical theory of stellar systems is not yet sufficiently developed. Although at first almost all the main points of the theory seemed to be verified (the Maxwellian distribution of velocities, the formation of an isothermal core, etc.) it is now realized that many problems need reconsideration.

There are many problems in this field that could be studied by means of numerical integrations.

(a) It would be interesting to increase the number of stars in a cluster. Such calculations need of course the use of faster electronic computers. However, one improvement in the method of integration, which may give good results in a relatively short time is the use of different integration steps for different orbits, according to the closeness of each star with the others.

(b) It would be very important to calculate the times of relaxation for different kinds of orbits, namely radial orbits and circular orbits. Further, one should calculate the evolution of a cluster whose orbits are initially approximately radial, or circular, or specific mixtures of these two types.

(c) One could study the distribution of the velocities at each point to find the change of the velocity ellipsoid. It should be useful to study also the distribution of the energy. Preliminary results (see [10]) indicate a distribution law $[-E]^{-\rho}$ with ρ variable in time. The distribution of the angular momentum is also important.

(d) The study of the formation and dissolution of binaries is another important topic. Are these binaries similar to the usual binary stars? It seems that triple or quadruple stars can also be formed.

(e) It would be interesting to study not only spherical, but elliptical distributions of stars also.

(f) Another relatively easy problem is the study of an expanding system, like an association. The other extreme, that of a collapsing system could perhaps explain some phases of the evolution of a stellar system.

Of special importance is the study of the final evolution of a stellar system, after most of its stars have escaped. It seems that this final state is a multiple or even a double star. This problem would need a very long calculating time. However, even the calculation with a very small number of stars, say 8 or even 4, would give useful results.

(g) Another problem arises if we drop the usual assumption that all the stars have equal mass. The distribution of stars of different masses is also very important.

(h) The evolution of a system whose stars change mass is still another problem. Thus one could explain the distribution of red giants, novae, and white dwarfs in clusters and in the Galaxy. Such a problem was considered lately by Boersma [46] and Blaauw [47]. Blaauw explains the big velocities of the so-called run-away stars by assuming that they belonged to binaries whose primaries lost mass very rapidly, so that the gravitational attraction suddenly diminished and the companions were ejected.

(i) Another problem is the evolution of a stellar system that is not isolated, but is subject to the tidal force of the Galaxy or of passing-by clouds. As a good approximation one should take the tidal field as due to just a point mass.

More problems of this type are considered by Ulam [45] and von Hörner [9], [10].

These problems can be studied by using fast numerical computers. At the present time the analytical methods in the n-body problem seem to have exhausted their power. However, there is a revival of interest in the n-body problem by using the results of numerical computations. These computations will give the necessary indications for further analytical progress, as well as for further numerical experiments. The fact that some of the first numerical results by von Hörner [9] have been already revised in [10] indicates that this field of research is still quite open.

III. **The continuum approach.**

1. *The Distribution Function.* The continuum approach considers a stellar system as composed of a very great number of particles distributed in a continuous way in space while their velocities form

also a continuous distribution in the velocity space. The points of the 6-dimensional phase space (the μ-space) are distributed as follows: If dN is the number of stars with coordinates between x_1 and $x_1 + dx_1$, x_2 and $x_2 + dx_2$, x_3 and $x_3 + dx_3$ and velocities between $\dot{x}_1$ and $\dot{x}_1 + d\dot{x}_1$, $\dot{x}_2$ and $\dot{x}_2 + d\dot{x}_2$, $\dot{x}_3$ and $\dot{x}_3 + d\dot{x}_3$, then

$$(37) \qquad dN = f(x_1, x_2, x_3, \dot{x}_1, \dot{x}_2, \dot{x}_3, t) \, dx_1 dx_2 dx_3 d\dot{x}_1 d\dot{x}_2 d\dot{x}_3,$$

where f is the distribution function. We write

$$(38) \qquad\qquad dN = f(x, \dot{x}, t) \, dx \, d\dot{x}.$$

The equations of motion are

$$(39) \qquad\qquad \frac{dx}{dt} = \dot{x}, \quad \frac{d\dot{x}}{dt} = -\frac{\partial V}{\partial x},$$

where V is the potential energy of a star per unit mass.

We shall prove that f is an integral of the equations of motion. In §II.4 we have seen that

$$(40) \qquad\qquad \frac{D(x, \dot{x})}{D(x_0, \dot{x}_0)} = 1.$$

On the other hand if we represent the number of stars in the six dimensional volume M_0 by $\iint_{M_0} f(x_0, \dot{x}_0, 0) \, dx_0 d\dot{x}_0$, this number remains the same during the evolution of the system; i.e., if $x = x(x_0, \dot{x}_0, t)$, $\dot{x} = \dot{x}(x_0, \dot{x}_0, t)$ is a solution of the equations of motion with initial conditions $x_0, \dot{x}_0$ for $t = 0$, then we have

$$(41) \qquad \iint_M f(x, \dot{x}, t) \, dx \, d\dot{x} = \iint_{M_0} f(x_0, \dot{x}_0, 0) \, dx_0 d\dot{x}_0.$$

But

$$\iint_M f(x, x, t) \, dx \, d\dot{x} = \iint_{M_0} f(x(x_0, \dot{x}_0, t), \dot{x}(x_0, \dot{x}_0, t), t) \frac{D(x, \dot{x})}{D(x_0, \dot{x}_0)} dx_0 d\dot{x}_0$$

$$= \iint_{M_0} f(x_0, x_0, 0) \, dx_0 d\dot{x}_0.$$

Hence by (40)

$$(42) \qquad \iint_{M_0} [f(x(x_0, \dot{x}_0, t), \dot{x}(x_0, \dot{x}_0, t), t) - f(x_0, \dot{x}_0, 0)] dx_0 d\dot{x}_0 = 0$$

for all regions M_0, i.e., $f(x, \dot{x}, t) = f(x_0, \dot{x}_0, 0)$ if $(x, \dot{x})$ is a solution of equations (39). Therefore f is a first integral of the equations of motion.

By differentiating f with respect to t we have

$$(43) \qquad \frac{\partial f}{\partial t} + \sum_{i=1}^{3} \left[\frac{\partial f}{\partial x_i} \dot{x}_i - \frac{\partial f}{\partial \dot{x}_i} \frac{\partial V}{\partial x_i} \right] = 0.$$

This is Liouville's equation for a stellar system.

The corresponding system to the partial differential equation (43) is

$$(44) \qquad dt = \frac{dx_1}{\dot{x}_1} = \frac{dx_2}{\dot{x}_2} = \frac{dx_3}{\dot{x}_3} = \frac{d\dot{x}_1}{-\dfrac{\partial V}{\partial x_1}} = \frac{d\dot{x}_2}{-\dfrac{\partial V}{\partial x_2}} = \frac{d\dot{x}_3}{\dfrac{\partial V}{\partial x_3}},$$

i.e., it is the same as the system (39).

If $I_1, I_2, \cdots, I_6$ are the 6 integrals of this system, the solution of (43) is $f = f(I_1, I_2, \cdots, I_6)$. This theorem is sometimes called Jeans' theorem [48] but it was given earlier by Poincaré (see [49], p. 101), and in any case it is a well-known theorem in the theory of differential equations. Two integrals of Liouville's equation are well known, namely the energy E and the angular momentum C.

If we consider only the distribution of the velocities in a region M_v near the point (x_1, x_2, x_3), we may write

$$(45) \qquad dN = F d\dot{x},$$

where

$$(46) \qquad F = \int_{M_v} f \, dx.$$

In order to describe the observed distribution of stellar velocities near the sun, K. Schwarzschild has given F in the form

$$(47) \qquad F = A \exp(- K^2 R^2 - H^2(\overline{\Theta}^2 + Z^2)),$$

where $A = NKH^2/\pi^{3/2}$, N is the number of stars in the region M_v, K and H are functions of the position only, while R is the radial velocity directed from the center outwards, $\overline{\Theta}$ is perpendicular to it on the galactic plane, and Z is perpendicular to the galactic plane.

Generalizations of this distribution are due to Charlier, Eddington, Jeans, Oort and others.

The most general ellipsoidal law is given by Chandrasekhar [50], [51], [2]. He takes

$$(48) \qquad f = A\phi \left(\sum_{i=1}^{3} \sum_{j=1}^{3} a_{ij}(\dot{x}_i - \dot{x}_{i0})(\dot{x}_j - \dot{x}_{j0}) + \sigma \right),$$

where $a_{ij}(= a_{ji})$, $\dot{x}_{10}, \dot{x}_{20}, \dot{x}_{30}$, σ and V are 11 unknown functions of the position and time only, such as to satisfy Liouville's equation. The velocity of the local standard of rest, i.e., the mean absolute velocity of the stars of the neighborhood of the point (x_1, x_2, x_3) is $(\dot{x}_{10}, \dot{x}_{20}, \dot{x}_{30})$.

If this formula is introduced into Liouville's equation, we find a system of 20 differential equations for the 11 unknown functions above. Chandrasekhar has solved completely the corresponding two-dimensional problem and many special cases of the three-dimensional problem also. He proved that a finite system in a steady state must necessarily have axial symmetry. The most general solution, however, has not been given.

Chandrasekhar studied in more detail the case when the velocity ellipsoid is reduced to a sphere:

$$(49) \qquad f = A\phi(a((\dot{x}_1 - \dot{x}_{10})^2 + (\dot{x}_2 - \dot{x}_{20})^2 + (\dot{x}_3 - \dot{x}_{30})^2) + \sigma).$$

The solutions include many spiral forms of the density distribution. The same spiral solutions are given by Schürer [52], by a much simpler method, namely by a change of coordinates. It was thought for some time that these solutions could explain the spiral form of the galactic system. However the spherical distribution of the residual velocities is not realized in the actual Galaxy. Further it must be stressed that even the general ellipsoidal distribution of the velocities is only an approximation. In fact ϕ should have not only quadratic terms in the residual velocities, but third order terms also etc. These third and higher terms are small if the velocities are small, and for this reason they are omitted in a first approximation.

Another limitation of Chandrasekhar's analysis is that it does not deal with self-gravitating stellar systems.

In a self-gravitating stellar system we have also Poisson's equation

(50) $$\Delta V \equiv \frac{\partial^2 V}{\partial x_1^2} + \frac{\partial^2 V}{\partial x_2^2} + \frac{\partial^2 V}{\partial x_3^2} = 4\pi G\rho,$$

where

(51) $$\rho = \int_{-\infty}^{\infty} F d\dot{x}$$

is the particle density. If stars of unequal masses are considered, then F is replaced[4] by $\int m\, F dm$; ρ is then the mass density.

By applying Poisson's equation one finds that not all the solutions of Liouville's equation represent self-gravitating stellar systems. E.g., Kurth [55], [56] proved that in the special case discussed by Schürer the system either is in a steady state or it is homogeneous. This led to the assumption that all ellipsoidal solutions of Liouville's equation are either in a steady state or homogeneous. This has caused a decrease of interest in these problems in the last years. However the problem is not yet settled completely. First there are some solutions that cannot be given by Schürer's transformation. Further the general ellipsoidal solution of Liouville's equation is not given. It would be useful to complete the discussion and see to what extent Kurth's theorem can be applied.

At any case there always remains the main problem of finding solutions that satisfy both Liouville's and Poisson's equations, especially for systems in a steady state. Prendergast [57] has given the general solution in the one-dimensional case. Other special solutions are given by Camm [53], [58] for stratified and spherical systems, by Kurth [59] for spherical systems, etc.

In the case of spherical stellar systems Poisson's equation becomes

(52) $$\frac{d^2 V}{dr^2} + \frac{2}{r}\frac{dV}{dr} = \frac{1}{r^2}\frac{d}{dr}\left(r^2\frac{dV}{dr}\right) = 4\pi G\rho.$$

Hence

(53) $$r^2\frac{dV}{dr} = 4\pi G \int_0^r \rho r^2 dr = Gm(r),$$

where $m(r)$ is the mass included in a sphere of radius r.

If the mass of a system is finite, then

---

[4] Cf. G. L. Camm [53]. A similar method was used earlier by J. Jeans [54, p. 231].

(54)
$$\frac{dV}{dr} = \frac{Gm(r)}{r^2} \leq \frac{Gm_\infty}{r^2},$$

and

(55) $$V - V(r_0) = G \int_{r_0}^r \frac{m(r)}{r^2} \, dr \leq Gm_\infty \int_{r_0}^r \frac{dr}{r^2} = Gm_\infty \left(\frac{1}{r_0} - \frac{1}{r} \right),$$

i.e.,

(56)
$$V_\infty \leq \frac{Gm_\infty}{r_0} + V(r_0).$$

Therefore, since V is an increasing function of r, it has a finite upper bound.

If $F = F(E, C)$ is a function of the two known integrals of Liouville's equation, then Kurth proved that for $E \geq V_\infty$, $F(E, C) = 0$.

Camm has considered the case where

(57)
$$F = \begin{cases} A(2E_1 - 2E)^\mu, & \text{if } E \leq E_1, \\ 0, & \text{if } E > E_1, \end{cases}$$

and $\mu > 0$.

This is consistent with the above restriction if $E_1 \leq V_\infty$. Then

(58)
$$\rho = \int \int \int_{-\infty}^{\infty} F dR \, d\bar{\Theta} dZ$$
$$= 2\pi A \int \int (-2V + 2E_1 - R^2 - S^2)^\mu dRSdS,$$

where $S = (\bar{\Theta}^2 + Z^2)^{1/2}$ is the transverse velocity, and the double integral extends to all the values where $S > 0$ and

(59) $$R^2 + S^2 \leq 2E_1 - 2V.$$

If we write $2E_1 - 2V = \Omega^2$ we find

(60)
$$\rho = 2\pi A \int_0^\Omega dR \int_0^{(\Omega^2 - R^2)^{1/2}} ((\Omega^2 - R^2) - S^2)^\mu dS^2$$
$$= \frac{2\pi A}{\mu + 1} \int_0^\Omega (\Omega^2 - R^2)^{\mu+1} dR,$$

and if we set $R = \Omega\sqrt{t}$

(61)
$$\rho = \frac{\pi A}{\mu + 1} \Omega^{2\mu+3} \int_0^1 (1 - t)^{\mu+1} t^{-1/2} dt$$
$$= \frac{\pi A}{\mu + 1} (2E_1 - 2V)^{\mu+3/2} B\left(\mu + 2, \frac{1}{2}\right)$$

where

$$B\left(\mu + 2, \frac{1}{2}\right) = \frac{\Gamma(\mu + 2)\,\Gamma\left(\frac{1}{2}\right)}{\Gamma\left(\mu + \frac{5}{2}\right)},$$

whence finally

(62)
$$\rho = \frac{\pi^{3/2} A\,\Gamma(\mu + 1)}{\Gamma\left(\mu + \frac{5}{2}\right)} (2E_1 - 2V)^{\mu+3/2}.$$

If we put this value in Poisson's equation (52), setting

(63) $\quad \Omega = -2V + 2E_1 = (-2V_0 + 2E_1)\psi \quad$ and $\quad r = cx$

with

(64) $\quad c^2 = \Gamma\left(\mu + \frac{5}{2}\right)(\Gamma(\mu + 1))^{-1}(8AG)^{-1}\pi^{-5/2}(2E_1 - 2V_0)^{-\mu-1/2},$

we get

(65)
$$\frac{1}{x^2}\frac{d}{dx}\left(x^2\frac{d\psi}{dx}\right) + \psi^{\mu+3/2} = 0.$$

This is the Emden equation of a polytropic gas sphere[5] with index $n = \mu + 3/2$. The initial conditions are $\psi = 1$ and $d\psi/dx = 0$ for $x = 0$ (because $dV/dr = 0$ for $r = 0$).

Emden [60] proved that for $n < 5$ we have finite radius and mass. For $n > 5$ the radius and mass are infinite. For $n = 5$ we have the solution

---

[5] A polytropic gas sphere is a sphere of gas whose pressure is proportional to the γ power of the density; $p \propto \rho^\gamma$. In Equation (65) we have $\gamma = 1 + 1/n$ and $\rho = \psi^n$.

(66)
$$\psi = \frac{(3q)^{1/2}}{(3 + q^2 x^2)^{1/2}},$$

and

(67)
$$\rho = \psi^5 = \frac{q^{5/2}}{\left(1 + \dfrac{q^2 x^2}{3}\right)^{5/2}} = \frac{B}{(a^2 + r^2)^{5/2}},$$

with

(68)
$$a^2 = \frac{3c^2}{q^2}, \quad \text{and} \quad B = \left(\frac{3c^2}{q}\right)^{5/2}.$$

Then the total mass $M = 4\pi B/3a^2$ is finite but the radius is infinite. This is the law of Schuster, applied by Plummer in the globular clusters. (See applications of polytropes [61], [62], [63].)

Another important case is that of an isothermal gas sphere. In this case we have a Boltzmann distribution of energy

(69)
$$F = A \exp(-2a^2 E),$$

and correspondingly a Maxwellian distribution of velocities. Then

(70)
$$\rho = A \int\int\int_{-\infty}^{\infty} \exp(-a^2(\dot{x}_1^2 + \dot{x}_2^2 + \dot{x}_3^2 + 2V)) \cdot d\dot{x}_1 d\dot{x}_2 d\dot{x}_3$$
$$= \frac{A\pi^{3/2}}{a^3} \exp(-2a^2 V),$$

or

(71)
$$\rho = \rho_0 \exp(-2a^2(V - V_0))$$

with

(72)
$$\rho_0 = \frac{A\pi^{3/2}}{a^3} \exp(-2a^2 V_0).$$

Then the equation of Poisson gives

(73)
$$\frac{1}{x^2} \frac{d}{dx}\left(x^2 \frac{du}{dx}\right) = \exp(-u),$$

with

(74)
$$2a^2(V - V_0) = u, \qquad r = cx, \qquad c^2 = (8\pi a^2 G \rho_0)^{-1}.$$

The initial conditions for $x = 0$ are $u = 0$ and $du/dt = 0$.

This is the equation of an isothermal gas sphere. This model gives an infinite radius and mass, as can be derived, e.g., from the fact that F does not become zero for any value of E. However it is assumed that this model represents approximately the central parts of stellar clusters.

It is proved that if the mass of a system is finite, then the condition $F(E, C) = 0$ for $E \geq V_\infty$ is necessary. This condition, however, is not sufficient. E.g., if we have a polytropic sphere with an index $n > 5$ the mass is infinite, although the condition is satisfied.

However Kurth proves [59] that in these cases there exists a function $\bar{F}(E, C)$ arbitrarily close to a given function $F(E, C)$ for which the corresponding mass and radius are finite.

Axially symmetrical stellar systems that are consistent with Poisson's equation have been considered by Camm [64], Kurth [55], Fricke [65], [66] and lately by Lynden-Bell [67].

It is an open question whether there exist systems where the distribution of velocities is everywhere ellipsoidal. Camm and Fricke proved under certain assumptions that no such system exists, while Kurth gave a method to obtain such systems. All these discussions are not yet complete [68]. Camm assumed that if a system has a finite mass then the potential has an asymptotic expansion

$$(75) \qquad V = -\frac{GM}{r} - \frac{S_c}{r^c} + \cdots$$

for large r (where $c > 1$), and in this case he proved that the density must be zero. However if

$$(76) \qquad V = -\frac{GM}{r}\left(1 - \frac{S}{\log r}\right),$$

for large r we cannot have an asymptotic expansion of the form (75) because

$$\left(\frac{1}{\log r}\right) \bigg/ \left(\frac{1}{r^{c-1}}\right) = \frac{r^{c-1}}{\log r} \longrightarrow \infty \quad \text{for } r \to \infty.$$

But the system (76), although it extends to infinity, has a finite mass, because Poisson's equation gives

$$\rho = \frac{1}{4\pi Gr^2}\frac{d}{dr}\left\{ r^2\frac{d}{dr}\left(-\frac{GM}{r}\left(1-\frac{S}{\log r}\right)\right)\right\}$$

(77)

$$= \frac{MS}{4\pi r^3}\left[\frac{1}{(\log r)^2}+\frac{2}{(\log r)^3}\right],$$

$$m_\infty - m(r_0) = MS\int_{r_0}^\infty\left(\frac{dr}{r(\log r)^2}+\frac{2dr}{r(\log r)^3}\right)$$

(78)

$$= MS\left[\frac{1}{\log r_0}+\frac{1}{(\log r_0)^2}\right].$$

Therefore there may be cases that are not covered by Camm's proof. On the other hand Kurth's solutions do not give a finite mass; in fact Kurth gives the potential function by solving formally Poisson's equation, written in the form

(79) $$\Delta V = 4\pi G\rho(V; x_1, x_2, x_3),$$

where

(80) $$\rho = \int\!\!\int\!\!\int_{-\infty}^{\infty} F(E, \cdots)\, d\dot x_1 d\dot x_2 d\dot x_3;$$

here $E = (\dot x_1^2 + \dot x_2^2 + \dot x_3^2)/2 + V$, and the dots after E represent the other integrals of motion. Then we can write

(81) $$V(x_1, x_2, x_3) = G\int\!\!\int\!\!\int_{-\infty}^{\infty} \frac{\rho(V(\xi_1, \xi_2, \xi_3); \xi_1, \xi_2, \xi_3)\, d\xi_1 d\xi_2 d\xi_3}{|x - \xi|}.$$

Kurth finds V either as a series in G, or $1/\sqrt{G}$, or by an iteration method, setting $V_0 = 0$ and computing V_{n+1}, after V_n is found, by the equation

(82) $$V_{n+1} = G\int\!\!\int\!\!\int_{-\infty}^{\infty} \frac{\rho(V_n; \xi_1, \xi_2, \xi_3)}{|x - \xi|}\, d\xi_1 d\xi_2 d\xi_3.$$

Then

(83) $$V = \lim_{n\to\infty} V_n.$$

Let us consider the last method and assume that there are only two isolating integrals E and C.

In a system of finite mass we may take $V_\infty = 0$; thus $\rho(0; \xi_1, \xi_2, \xi_3) = 0$ for all the values of C, because $F(E, C) = 0$ for $E = (\dot{x}_1^2 + \dot{x}_2^2 + \dot{x}_3^2)/2 > V_\infty = 0$. Therefore $V_1 = 0$ and generally $V_n = 0$ and $V = 0$, i.e., this method does not give V.

The same is the case with the series expansions of the form

$$(84) \qquad V = V_0(x_1, x_2, x_3) + GV_1(x_1, x_2, x_3) + \cdots,$$

and

$$(85) \qquad \rho = \rho_0(x_1, x_2, x_3) + G\rho_1(x_1, x_2, x_3) + \cdots,$$

because again for $V_0 = 0$ we have $\rho = \rho_0 = 0$ and then Equation (81) gives, step by step

$$V_1 = V_2 = \cdots = 0.$$

Therefore Kurth's method does not work for systems with finite mass.

Danby proved that in the case of an ellipsoidal distribution of velocities there is no third integral besides the energy and the angular momentum.

If we drop the ellipsoidal hypothesis we can find axially symmetrical systems whose distribution function F is a function of E and C. In fact Lynden-Bell [67] solves the equation

$$\rho = \int_{-\infty}^{\infty} F d\dot{x}$$

for axially symmetric systems when ρ is given, and finds $F = F(E, C)$. There is a special solution that is even with respect to C; i.e., $F = F(E, C^2)$.

The general solution contains also any odd function ΔF with respect to C because

$$(86) \qquad \int_{-\infty}^{\infty} \Delta F d\dot{x} = 0.$$

ΔF is found by Lynden-Bell by means of relaxation considerations, assuming that the stellar encounters cause the velocity dispersions along the 3 axes to be equal. This, however, is not the case in our Galaxy. (See also [236].)

The distribution function of a spherical system is not necessarily a function of C only, as was assumed by Jeans [69, p. 365]. The general form of F is

(87) $$F = F(E, C_1, C_2, C_3),$$

where C_1, C_2, C_3 are the three components of the total angular momentum.

Lynden-Bell [70] and Woolley [71] considered a case where a spherical cluster is rotating. Such a rotating cluster can be constructed out of a nonrotating one by inverting the velocities of the stars moving, say, clockwise with respect to a given axis.

More generally, we may invert the different velocities of different groups of stars rotating around different axes, without changing the spherical form of the cluster.

Fricke [65], [66] has made a severe criticism of the ellipsoidal theory of stellar velocities. He points out that the ellipsoidal distribution is only a rough approximation to the real distribution of stellar velocities, and then only if these velocities are small. It is well known that the high velocity stars have a very asymmetric distribution and therefore their motions can by no means be described by the ellipsoidal theory.

Fricke made a search for more general distribution functions F of the form $F(E, C)$ that are not quadratic in the velocities, but satisfy Poisson's equation. He considered integrals of the form

$$F = \sum_{i,k} a_{ik} E^i C^k$$

(polynomials in E and C) for systems with finite mass. The solutions found are of the form

(88) $$F = E^{5/2} \sum A_{kl} E^{k+l} C^{2k} + E^{N+7/2} \sum B_{kl} E^{k+l} C^{2k+1}.$$

Here N is a nonnegative integer and in both summations k goes from 0 to n and l goes from 0 to m. Fricke gives the examples

(89) $$F = (2E)^{5/2}(2 + 100(2E)^4 C^6 - 49(2E)^3 C^5(1 + (2E)^2 C^2)),$$

and

(90) $$F = (2E)^{5/2}(0.5 + 100(2E)^{11} C^{20} + 49.5(2E)^{10} C^{19}(1 + (2E)^2 C^2));$$

these two distributions of velocities can describe both the almost ellipsoidal distribution of small velocities and the big asymmetric velocities of the high velocity stars.

However the functions used by Fricke seem rather artificial. No obvious reason suggests the use of any of the above functions F.

Further, the fact that the velocity ellipsoid is flattened along the z-axis cannot be explained if only two integrals of motion E and C are used in the distribution function F.

2. *Integrals of Motion.* The equations of motion in an axially symmetrical potential field are

$$(91) \qquad \frac{dr}{R} = \frac{dz}{Z} = \frac{dR}{-\dfrac{\partial V}{\partial r} + \dfrac{\Theta^2}{r}} = \frac{d\Theta}{-\dfrac{R\Theta}{r}} = \frac{dZ}{-\dfrac{\partial V}{\partial z}} = dt,$$

and these have two well-known integrals: the energy

$$E = (R^2 + \Theta^2 + Z^2)/2 + V,$$

and the angular momentum $C = r\Theta$.

The general form of an integral that is a function of E and C and is quadratic in the velocities, is

$$(92) \qquad I = 2E - 2k_1 C + k_2 C^2,$$

or

$$(93) \qquad \begin{aligned} I &= R^2 + Z^2 + \Theta^2(1 + k_2 r^2) - 2k_1 r\Theta + 2V \\ &= R^2 + Z^2 + (1 + k_2 r^2)\overline{\Theta}^2 + 2V - \frac{k_1^2 r^2}{1 + k_2 r^2}, \end{aligned}$$

where

$$(94) \qquad \overline{\Theta} = \Theta - \Theta_0 \quad \text{and} \quad \Theta_0 = \frac{k_1 r}{1 + k_2 r^2} \ (^6).$$

(^{6}) Then the constants of Oort are

$$A = \frac{1}{2}\left(\frac{\Theta_0}{r} - \frac{d\Theta_0}{dr}\right) = \frac{k_1 k_2 r^2}{(1 + k_2 r^2)^2},$$

and

$$B = -\frac{1}{2}\left(\frac{\Theta_0}{r} + \frac{d\Theta_0}{dr}\right) = -\frac{k_1}{(1 + k_2 r^2)^2};$$

$$\omega = \frac{\Theta_0}{r} = A - B = \frac{k_1}{1 + k_2 r^2},$$

and the square of the ratio of the axes b (along $\overline{\Theta}$) and a (along R) is

$$\frac{b^2}{a^2} = \frac{1}{1 + k_2 r^2} = 1 - \frac{A}{\omega}.$$

If $k_2 = 0$ we have three equal axes. This happens when statistical equilibrium is reached. In this case $\theta_0 = k_1 r$, i.e., the system rotates like a rigid body.

The velocity ellipsoid has the axes R and Z equal. However in K. Schwarzschild's ellipsoidal distribution defined by (47) the axes $\bar{\theta}$ and Z are equal, and much smaller than R. In a more general form

$$(95) \qquad F = A \exp \left(-\frac{R^2}{\sigma_1^2} - \frac{\bar{\theta}^2}{\sigma_2^2} - \frac{Z^2}{\sigma_3^2} \right) ;$$

the values of the dispersions σ_1, σ_2, σ_3 are given for different groups of stars by Nordstöm [72]. In general it is $\sigma_1 > \sigma_2 > \sigma_3$, but sometimes $\sigma_1 > \sigma_3 > \sigma_2$. In any case the axis R is always greater than the other axes.

The observed inequality of the R and Z axes cannot be explained if only two integrals of the equations of motion are used.

As we have seen, Liouville's equation has 6 integrals of motion that are, in general, time dependent. By elimination of the time we get 5 time independent integrals. These, however, are not in general isolating (Wintner [73, p. 93-97]; Contopoulos [74]); i.e., if they are solved for one variable as a function of the others they give an infinity of solutions that are densely distributed in some interval. E.g., in the case of two perpendicular oscillations we have the system

$$(96) \qquad \frac{dx}{X} = \frac{dy}{Y} = \frac{dX}{-Ax} = \frac{dY}{-By} = dt,$$

(X, Y are the components of the velocity), with the integrals

$$(97) \qquad \Phi_{10} = \frac{1}{2} \left(X^2 + Ax^2 \right), \quad \Phi_{20} = \frac{1}{2} \left(Y^2 + By^2 \right).$$

If $A^{1/2}/B^{1/2}$ is irrational, the third time-independent integral

$$(98) \qquad T_0 = \frac{1}{B^{1/2}} \sin^{-1} \frac{B^{1/2} y}{(2\Phi_{20})^{1/2}} - \frac{1}{A^{1/2}} \sin^{-1} \frac{A^{1/2} x}{(2\Phi_{10})^{1/2}},$$

is not isolating.

In fact we have

$$(99) \qquad \frac{B^{1/2} y}{(2\Phi_{20})^{1/2}} = \sin \left(B^{1/2} T_0 + \frac{B^{1/2}}{A^{1/2}} \sin^{-1} \frac{A^{1/2} x}{(2\Phi_{10})^{1/2}} \right),$$

and for a given value of x we have infinitely many values

$$\sin^{-1}\frac{A^{1/2}x}{(2\Phi_{10})^{1/2}} = s_1 + 2K\pi \quad \text{or} \quad \sin^{-1}\frac{A^{1/2}x}{(2\Phi_{10})^{1/2}} = (2K+1)\pi - s_1.$$

Thus

$$\sin\left(B^{1/2}T_0 + \frac{B^{1/2}}{A^{1/2}}(s_1 + 2K\pi)\right)$$

$$= \sin\left(B^{1/2}\left(T_0 + \frac{s_1}{A^{1/2}}\right) + \frac{2K\pi B^{1/2}}{A^{1/2}} - 2L\pi\right),$$

where K, L are arbitrary integers; $2\pi(K(B^{1/2}/A^{1/2}) - L)$ can be made to approach any number arbitrarily closely, by a suitable choice of the integers K, L. Therefore the values of $B^{1/2}y/(2\Phi_{20})^{1/2}$ are densely distributed in the interval $(-1, 1)$ and this means that the integral is not isolating. If $A^{1/2}/B^{1/2}$ is rational, we have only a finite number of values of y for any value of x and the integral T_0 is then isolating.

Many authors have used isolating integrals besides the energy and the angular momentum in some special cases. E.g., B. Lindblad [75], introduced the integral

$$(100) \qquad I_3 = Z^2 + 2(V - V_0),$$

where $V_0 = V(r, 0)$. In this case, however $(\partial^2 V/\partial r\,\partial z) = 0$; this integral is approximately valid for very flattened systems. Van Albada [76] and Kuzmin [77], [78], [79] have studied especially the integral

$$(101) \qquad I_3 = (rZ - zR)^2 + z^2\Theta^2 + c_0^2(Z^2 + 2V^*),$$

where V^* is connected by two relations to the potential energy V, and c is a constant. This integral was introduced for the first time by Eddington [80]. In this case V is of the form

$$(102) \qquad V = -\left[\frac{F(\xi_1) - G(\eta_1)}{\xi_1 - \eta_1}\right]$$

where

$$(103) \qquad \begin{array}{l}\xi_1\\\eta_1\end{array} = r^2 + z^2 \pm (c^4 - 2c^2(z^2 - r^2) + (r^2 + z^2)^2)^{1/2}.$$

A general case where an isolating third integral can be found has been given by Stäckel [81], [82]. In the two-dimensional case

$$(104) \qquad V = - \left[\frac{M(\xi) + N(\eta)}{K(\xi) + L(\eta)} \right],$$

while the line element is

$$(105) \qquad ds^2 = (K(\xi) + L(\eta))(d\xi^2 + d\eta^2).$$

In this case the corresponding momenta to the variables ξ, η are

$$(106) \qquad p_\xi = (K(\xi) + L(\eta))\dot\xi, \qquad p_\eta = (K(\xi) + L(\eta))\dot\eta,$$

hence the energy integral is

$$(107) \qquad \frac{1}{2} \left[\frac{p_\xi^2 + p_\eta^2}{K(\xi) + L(\eta)} \right] - \left[\frac{M(\xi) + N(\eta)}{K(\xi) + L(\eta)} \right] = a.$$

Then the Hamilton-Jacobi equation is separable, and two integrals can be easily found

$$(108) \qquad \begin{aligned} (K(\xi) + L(\eta))^2 \dot\xi^2/2 - M(\xi) - aK(\xi) &= -b, \\ (K(\xi) + L(\eta))^2 \dot\eta^2/2 - N(\eta) - aL(\eta) &= b. \end{aligned}$$

By adding these two integrals we derive the energy integral (107). Stäckel considered also systems of more than two dimensions, where we have separation of variables in the Hamilton-Jacobi equation.

Weinacht [83] proved that if the line element is Euclidean the most general cases where we have separation of variables in 2 or 3 dimensions which can be reduced to Cartesian coordinates by a point transformation, are of the Stäckel type.

Van de Hulst [84] used the transformation

$$(109) \qquad \begin{aligned} r &= c \sinh \xi \cos \eta - k + r_0, \\ z &= c \cosh \xi \sin \eta, \end{aligned}$$

and a potential of the form (104) with

$$K(\xi) = 2c^2 \cosh^2 \xi, \qquad L(\eta) = -2c^2 \sin^2 \eta,$$

$$M(\xi) = -2c^2 \cosh^2 \xi V_0, \qquad N(\eta) = -Q(c^2 + k^2) 2 \sin^2 \eta,$$

where c, k, r_0, Q are constants, and V_0 is the potential for $\eta = 0$.

This is a generalization of Eddington's potential, which corresponds to $k = r_0$. It is easily proved that if $k = r_0$

(110)
$$c^2(2\cosh^2\xi - 1) = \xi_1 \qquad (\text{or} = \eta_1),$$
$$c^2(2\sin^2\eta - 1) = \eta_1 \qquad (\text{or} = \xi_1).$$

Van de Hulst's transformation (109) is the most general point-transformation in two coordinates that gives separation of variables. More general transformations that give separation of variables one can find only by using contact transformations (Siegel [32], Contopoulos [74], [85]).

A systematic search and classification of potentials with isolating integrals has been done lately by Lynden-Bell [86], [87]. Lynden-Bell considers only integrals depending on an arbitrary function (corresponding to potentials that also depend on the same arbitrary function), and he calls such integrals "local."

In general, however, a dynamical problem is nonseparable, or more accurately, there is no known transformation that can reduce it to a separable one.

In such cases a third integral of motion can be introduced in the form of a series. Whittaker [88] and [89, Chapter 16], called this integral the "adelphic integral" (from the greek word "adelphos," meaning brother) because it is very similar to the energy integral. This is given in the form of a trigonometric series. Another form of this integral was given by Cherry [90], [91], [92]. The third integral in a form directly applicable to stellar systems was given by Contopoulos [93].

If we have a Hamiltonian system of the form $H = H_2 + H_3 + H_4 + \cdots$, where H_k is an homogeneous polynomial of degree k in the variables, we can make a linear canonical transformation of variables so that H_2 becomes

$$H_2 = \sum_{i=1}^n \lambda_i x_i y_i.$$

Then there is a formal canonical transformation of variables

(111)
$$\overline{x}_i = F(x_i, y_i) = x_i + \cdots,$$
$$\overline{y}_i = G(x_i, y_i) = y_i + \cdots,$$

such that H becomes a function of the products $v_i = \overline{x}_i \overline{y}_i$ only.

The canonical equations become

(112) $\quad \dfrac{d\bar{x}_i}{dt} = \dfrac{\partial H}{\partial \bar{y}_i} = m_i \bar{x}_i, \quad \dfrac{d\bar{y}_i}{dt} = -\dfrac{\partial H}{\partial \bar{x}_i} = -m_i \bar{y}_i \ (i = 1, 2, \cdots, n),$

where

(113) $\qquad\qquad\qquad\qquad m_i = \dfrac{\partial H}{\partial v_i}.$

Then

$$\bar{y}_i \frac{d\bar{x}_i}{dt} + \bar{x}_i \frac{d\bar{y}_i}{dt} = 0 \quad \text{or} \quad \frac{dv_i}{dt} = 0.$$

Hence we have n integrals

(114) $\qquad\qquad\qquad\qquad v_i = \bar{x}_i \bar{y}_i = \text{const.}$

Then m_i are constants, and therefore the solution of the Hamiltonian system is

(115) $\qquad\qquad \bar{x}_i = a_i \exp(m_i t), \quad \bar{y}_i = b_i \exp(m_i t).$

The same transformation was introduced by Birkhoff [**94**, Chapters 3, 8]. His method is similar to the methods of Delaunay [**95**, p. 541] and von Zeipel [**96**] in classical Celestial Mechanics (see also [**74**]).

We shall presently apply the von Zeipel method in the case of two dimensions.

If

(116) $\qquad\qquad\qquad\qquad H = H_2 + H_3 + \cdots,$

where

(117) $\qquad\qquad\qquad\qquad H_2 = \lambda_1 x_1 y_1 + \lambda_2 x_2 y_2,$

we introduce the determining function

(118) $\qquad\qquad S = S(x_1, x_2, \bar{y}_1, \bar{y}_2) = S_0 + S_1 + \cdots,$

where

(119) $\qquad\qquad\qquad\qquad S_0 = x_1 \bar{y}_1 + x_2 \bar{y}_2,$

such that the new Hamiltonian $\bar{H}$ is a function of $v_1 = \bar{x}_1 \bar{y}_1$ and $v_2 = \bar{x}_2 \bar{y}_2$ only. We have

(120) $\qquad y_i = \dfrac{\partial S}{\partial x_i} = \bar{y}_i + \dfrac{\partial S_1}{\partial x_i} + \cdots, \quad \bar{x}_i = x_i + \dfrac{\partial S_1}{\partial \bar{y}_i} + \cdots.$

Hence

$$x_i = \bar{x}_i - \frac{\partial S_1}{\partial \bar{y}_i} - \frac{\partial S_2}{\partial \bar{y}_i} - \sum_{j=1}^{2} \left\{ \frac{\partial^2 S_1}{\partial \bar{y}_i \partial \bar{x}_j} \left(-\frac{\partial S_1}{\partial \bar{y}_j} + \cdots \right) \right.$$

$$\left. + \frac{\partial^2 S_1}{\partial \bar{y}_i \partial \bar{y}_j} \left(\frac{\partial S_1}{\partial \bar{x}_j} + \cdots \right) \right\} + \cdots,$$

(121)

$$y_i = \bar{y}_i + \frac{\partial S_1}{\partial \bar{x}_i} + \frac{\partial S_2}{\partial \bar{x}_i} + \sum_{j=1}^{2} \left\{ \frac{\partial^2 S_1}{\partial \bar{x}_i \partial \bar{x}_j} \left(-\frac{\partial S_1}{\partial \bar{y}_j} + \cdots \right) \right.$$

$$\left. + \frac{\partial^2 S_1}{\partial \bar{x}_i \partial \bar{y}_j} \left(\frac{\partial S_1}{\partial \bar{x}_j} + \cdots \right) \right\} + \cdots,$$

where the derivatives are calculated for $\bar{x}_i$ and $\bar{y}_i$. Then

$$H(x_i, y_i) = H\left(\bar{x}_i - \frac{\partial S_1}{\partial \bar{y}_i} - \cdots, \ \bar{y}_i + \frac{\partial S_1}{\partial \bar{x}_i} + \cdots \right)$$

$$= H_2 - \sum_{i=1}^{2} \frac{\partial H_2}{\partial \bar{x}_i} \frac{\partial S_1}{\partial \bar{y}_i} + \sum_{i=1}^{2} \frac{\partial H_2}{\partial \bar{y}_i} \frac{\partial S_1}{\partial \bar{x}_i}$$

$$+ H_3 - \sum_{i=1}^{2} \frac{\partial H_3}{\partial \bar{x}_i} \frac{\partial S_1}{\partial \bar{y}_i} + \sum_{i=1}^{2} \frac{\partial H_3}{\partial \bar{y}_i} \frac{\partial S_1}{\partial \bar{x}_i}$$

(122) $\qquad - \sum_{i=1}^{2} \frac{\partial H_2}{\partial \bar{x}_i} \left[\frac{\partial S_2}{\partial \bar{y}_i} - \sum_{j=1}^{2} \left\{ \frac{\partial^2 S_1}{\partial \bar{y}_i \partial \bar{x}_j} \frac{\partial S_1}{\partial \bar{y}_j} - \frac{\partial^2 S_1}{\partial \bar{y}_i \partial \bar{y}_j} \frac{\partial S_1}{\partial \bar{x}_j} \right\} \right]$

$$+ \sum_{i=1}^{2} \frac{\partial H_2}{\partial \bar{y}_i} \left[\frac{\partial S_2}{\partial \bar{x}_i} - \sum_{j=1}^{2} \left\{ \frac{\partial^2 S_1}{\partial \bar{x}_i \partial \bar{x}_j} \frac{\partial S_1}{\partial \bar{y}_j} - \frac{\partial^2 S_1}{\partial \bar{x}_i \partial \bar{y}_j} \frac{\partial S_1}{\partial \bar{x}_j} \right\} \right]$$

$$+ \sum_{i=1}^{2} \sum_{j=1}^{2} \frac{1}{2} \left\{ \frac{\partial^2 H_2}{\partial \bar{x}_i \partial \bar{x}_j} \frac{\partial S_1}{\partial \bar{y}_i} \frac{\partial S_1}{\partial \bar{y}_j} - \frac{\partial^2 H_2}{\partial \bar{x}_i \partial \bar{y}_j} \frac{\partial S_1}{\partial \bar{y}_i} \frac{\partial S_1}{\partial \bar{x}_j} \right.$$

$$\left. + \frac{\partial^2 H}{\partial \bar{y}_i \partial \bar{y}_j} \frac{\partial S_1}{\partial \bar{x}_i} \frac{\partial S_2}{\partial \bar{x}_j} \right\}$$

$$+ H_4 + \cdots = \bar{H}_2 + \bar{H}_3 + \bar{H}_4 + \cdots;$$

the derivatives of $H_2, H_3, \cdots$ are also calculated for $\bar{x}_i, \bar{y}_i$. E.g.,

(123) $$\frac{\partial H_2}{\partial \bar{x}_i} = \lambda_i \bar{y}_i, \quad \frac{\partial H_2}{\partial \bar{y}_i} = \lambda_i \bar{x}_i, \text{ etc.}$$

Hence

(124) $$H_2 = \bar{H}_2,$$

(125) $$\sum_{i=1}^{2} \lambda_i \bar{x}_i \frac{\partial S_1}{\partial \bar{x}_i} - \sum_{i=1}^{2} \lambda_i \bar{y}_i \frac{\partial S_1}{\partial \bar{y}_i} = \bar{H}_3 - H_3 = Q_1,$$

and generally

(126) $$\sum_{i=1}^{2} \lambda_i \bar{x}_i \frac{\partial S_k}{\partial \bar{x}_i} - \sum_{i=1}^{2} \lambda_i \bar{y}_i \frac{\partial S_k}{\partial \bar{y}_i} = \bar{H}_k - Q'_k = Q_k.$$

where Q'_k is found after the terms $S_1, \cdots, S_{k-1}$ are defined. If we write $Q_k = \bar{H}_k - Q'_k$, and take $\bar{H}_k$ to be the sum of the terms of the form $(\bar{x}_1 \bar{y}_1)^{a_1} (\bar{x}_2 \bar{y}_2)^{a_2}$ in Q'_k, then $\bar{H}_k$ is a function of v_1 and v_2 only, and Q_k contains only terms of the form $\bar{x}_1^{a_1} \bar{y}_1^{b_1} \bar{x}_2^{a_2} \bar{y}_2^{b_2}$ in which $(a_1 - b_1)^2 + (a_2 - b_2)^2 \neq 0$.

The corresponding system to the Equation (126) is

(127) $$\frac{d\bar{x}_1}{\lambda_1 \bar{x}_1} = \frac{d\bar{x}_2}{\lambda_2 \bar{x}_2} = -\frac{d\bar{y}_1}{\lambda_1 \bar{y}_1} = -\frac{d\bar{y}_2}{\lambda_2 \bar{y}_2} = \frac{dS_k}{Q_k}$$

and it has the integrals

(128) $$v_1 = \bar{x}_1 \bar{y}_1, \quad v_2 = \bar{x}_2 \bar{y}_2 \quad \text{and} \quad \frac{\bar{x}_1^{1/\lambda_1}}{\bar{x}_2^{1/\lambda_2}} = c.$$

Therefore S_k is given by

(129) $$S_k = \int Q_k \frac{d\bar{x}_1}{\lambda_1 \bar{x}_1} + S_{k0}(v_1, v_2, c),$$

where in Q_k we have replaced $\bar{y}_1$, $\bar{y}_2$, by $v_1/\bar{x}_1$, $v_2/\bar{x}_2$ and $\bar{x}_2$ by $\bar{x}_1^{\lambda_2/\lambda_1} c^{-\lambda_2}$; S_{k0} is an arbitrary function of v_1, v_2, c.

Let $\bar{x}_1^{a_1} \bar{y}_1^{b_1} \bar{x}_2^{a_2} \bar{y}_2^{b_2}$ be a term of Q_k. This is written

$$v_1^{b_1} v_2^{b_2} \bar{x}_1^{\lambda/\lambda_1} c^{-\lambda_2(a_2 - b_2)}$$

where

$$\lambda = \lambda_1(a_1 - b_1) + \lambda_2(a_2 - b_2).$$

After integration it gives

$$v_1^{b_1} v_2^{b_2} \bar{x}_1^{\lambda/\lambda_1} c^{-\lambda_2(a_2-b_2)}\lambda^{-1},$$

which can be rewritten as

(130)
$$\frac{\bar{x}_1^{a_1}\bar{y}_1^{b_1}\bar{x}_2^{a_2}\bar{y}_2^{b_2}}{\lambda_1(a_1-b_1)+\lambda_2(a_2-b_2)}.$$

The denominator $(\lambda_1(a_1-b_1)+\lambda_2(a_2-b_2))$ is never zero if λ_1/λ_2 is irrational, because $(a_1-b_1)^2+(a_2-b_2)^2 \neq 0$.

Thus S and $\bar{H}$ are constructed step by step; both are functions of $\bar{x}_i$ and $\bar{y}_i$, and $\bar{H}=f(v_1,v_2)$. Then the transformations (121) give x_i, y_i as functions of $\bar{x}_i$, $\bar{y}_i$. The problem of the convergence of S and $\bar{H}$, however, is left open.

If λ_1/λ_2 is rational, one can find S by a convenient choice of the arbitrary functions $S_{k-j,0}$ $(0 < j)$, so that there should not appear terms of the form

$$\bar{x}_1^{a_1}\bar{y}_1^{b_1}\bar{x}_2^{a_2}\bar{y}_2^{b_2} \text{ in } Q_k \text{ with } \lambda_1(a_1-b_1)+\lambda_2(a_2-b_2)=0.$$

The elimination of such (secular) terms has been discussed by Whittaker [88], [89], Cherry [91], and Contopoulos [237].

In the case of a stellar system with an axis and a plane of symmetry we can develop the potential function in a region around a point $(r_0, z_0 = 0)$ in the plane of symmetry in the form of a series

(131) $2V = -(C^2/r^2 - P\xi^2 - Qz^2 + 2a\xi^3/3 + 2b\xi z^2 + \cdots),$

where $\xi = r - r_0$; C is the constant of areas and P, Q, a, b are constants.

In the case of our Galaxy one can take approximately $P = 0.076$ $(10^7 \text{ years})^{-2}$, $Q = 0.550$ $(10^7 \text{ years})^{-2}$, $a = 0.052$ $(10^7 \text{ years})^{-2} \text{ kpc}^1$, $b = 0.206$ $(10^7 \text{ years})^{-2} \text{ kpc}^{-1}$.

A third integral of motion can now be found as a formal series (see [93]):

(132) $\Phi = \Phi_0 + a\Phi_a + b\Phi_b + \cdots + a^2\Phi_{aa} + ab\Phi_{ab} + b^2\Phi_{bb} + \cdots$

where

(133)
$$\Phi_0 = \frac{1}{2}(P\xi^2 + R^2), \qquad \Phi_a = -\frac{1}{3}\xi^3,$$

$$\Phi_b = \frac{1}{4Q-P}((P-2Q)\xi z^2 - 2\xi Z^2 + 2RzZ),$$

etc. The accuracy of this integral is very great. In two orbits calculated for 5×10^9 years, the accuracy is better than $1/1000$ if terms up to the third degree are included (see [97]).

The existence of an isolating or quasi-isolating third integral can explain the three-axial form of the velocity ellipsoid. In fact, we may take approximately

(134)
$$\Phi = \frac{1}{2} (P\xi^2 + R^2) - \frac{a\xi^3}{3}$$
$$+ \frac{b}{4Q - P} ((P - 2Q)\xi z^2 - 2\xi Z^2 + 2RzZ),$$

and this is quadratic in the velocities. Then

(135)
$$F = A\phi(I)$$

where

(136)
$$I = 2E - 2k_1 C + k_2 C^2 + 2k_3 \Phi = R^2(1 + k_3) + (1 + k_2 r^2)(\Theta - \Theta_0)^2$$
$$+ Z^2 \left(1 - \frac{4k_3 b\xi}{4Q - P}\right) + \frac{4k_3 bz}{4Q - P} RZ - \frac{k_1^2 r^2}{1 + k_2 r^2}$$
$$+ k_3 \left(P\xi^2 - \frac{2a\xi^3}{3} + \frac{2b}{4Q - P} (P - 2Q)\xi z^2\right).$$

This represents a three-axial ellipsoid whose main axis forms a small angle (proportional to z) with the radial direction R. This angle is exactly zero if $z = 0$.

Barbanis [97] gives for the velocities R and Z the distribution

(137)
$$dN = A_1 e^{-2kE - 2l\Phi} dR dZ,$$

and he finds $k = 19.7$ (kpc/10^7 years)$^{-2}$, $l = 14.1$ (kpc/10^7 years)$^{-2}$. The axes Z and R in the meridian plane have a ratio 0.5, as given by observation (see [98]).

3. *Galactic Orbits.* The orbits of the stars in a stellar system are in general calculated without taking into account the effects of the encounters; i.e., a star is supposed to move in the general potential field of the stellar system, considered as a continuum, and any deviations from this motion, because of close encounters, are disregarded. This is because the time of relaxation is very great in comparison with the mean periods of the motions of the stars. E.g., in the Galaxy the time of relaxation is of the order of 10^{14}

years (see [2]), if only the interactions of stars are considered, while the period of galactic rotation is about 2×10^8 years. This means that a star may, under these conditions, describe 10^5 revolutions around the center of the Galaxy before its orbit is markedly changed([7]).

In a globular cluster the time of relaxation is about 10^9 to 10^{10} years (see [2]), while the rotation periods are of the order of 10^7 years or smaller. Again the stars may describe many revolutions before their orbits are markedly changed.

a. *Plane Orbits*. The orbits of stars in a globular cluster or in the plane of symmetry of a galaxy have been widely studied.

In the case of a globular cluster (see [99], [100], [101], [102]) the potential energy per unit mass is

(138)
$$V = - \frac{Gm(r)}{r} - 4\pi G \int_r^R r\rho(r)\, dr,$$

where $m(r)$ is the mass of the cluster inside a sphere of radius r, ρ is the density and R is the radius of the cluster (that may be considered infinite). The equations of motion are now

(139)
$$\frac{1}{2}(v^2 - v_0^2) = V(r_0) - V(r) \quad \text{and} \quad r^2 \frac{d\theta}{dt} = C.$$

If we set

(140)
$$w(r) = - 2G \int_0^r \frac{m(r)}{r^2}\, dr,$$

and

(141)
$$h = v_0^2 - w(r_0),$$

we get

(142)
$$v^2 = w(r) + h = \left(\frac{dr}{dt}\right)^2 + \frac{C^2}{r^2}.$$

Thus we find

(143)
$$\left(\frac{dr}{dt}\right)^2 = \frac{(w(r) + h)r^2 - C^2}{r^2} = \frac{\phi(r)}{r^2},$$

([7]) We shall see that other effects, besides the interactions of stars, make the time of relaxation much shorter. However, it is longer than one rotation period.

hence

(144)
$$t - t_0 = \int_{r_0}^{r} \frac{rdr}{(\phi(r))^{1/2}}.$$

Similarly

(145)
$$\theta - \theta_0 = \int_{r_0}^{r} \frac{Cdr}{r(\phi(r))^{1/2}},$$

where (r_0, θ_0) is the initial position and $(dr/dt)_0$ is the initial radial velocity for $t = t_0$.

The function $\phi(r)$ is positive or zero for r_0 (because $r_0^2(dr/dt)_0^2 \geqq 0$), and it is negative for $r = 0$ (if $C \neq 0$), i.e., there is a root r_1 be-

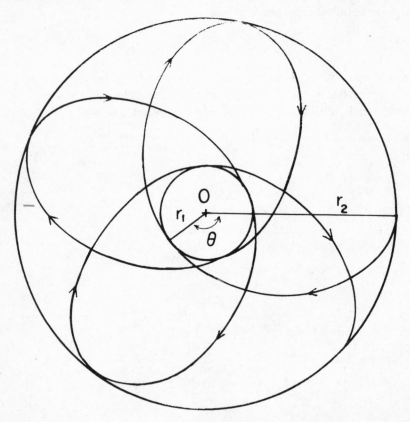

FIGURE 5. Angle from Apocentron to Pericentron

tween 0 and r_0. If $w(r) \to -\infty$ when $r \to \infty$, or if $w(\infty)$ is finite and $h < -w(\infty)$, then there is another root $r_2 \geqq r_0$.

It can be proved that only these two positive roots exist; if $h \geqq -w(\infty)$ no second root exists.

In the first case the orbits move between a minimum r_1 (pericentron) and a maximum r_2 (apocentron) (Figure 5) and the angle θ between the directions of the pericentron and apocentron is in general between $\pi/2$ and π (see [100]). If this angle is commensurable with π the orbit is periodic. If however θ/π is irrational, the orbit is not periodic and it "fills" densely the whole annulus between the circles r_1 and r_2, i.e., it comes arbitrarily near any point of this annulus. Such an orbit is called ergodic. An ergodic orbit is also recurrent, i.e., it returns an infinite number of times near the initial point.

In the second case the orbit extends to infinity. It is similar to a hyperbola or a parabola.

In the case of a motion in the plane of an axially symmetrical system some other possibilities may appear. In this case $\phi(r)$ may have a third root $r_3 \geqq r_2$ and either $r_1 \leqq r_0 \leqq r_2$ or $r_3 \leqq r_0$. If $r_1 \leqq r_0 \leqq r_2 < r_3$ and $r_1 < r_2$, the motion is represented by the rosette of of Figure 5. If $r_2 < r_3 \leqq r_0$, the motion extends to infinity. If, however, $r_2 = r_3$ then we have an exceptional case. The orbit issuing from r_0 and moving towards r_2 approaches asymptotically the circle r_2, along a spiral (Figure 6) but it does not reach it in a finite time. In fact then the integrals

$$\int_{r_0}^{r_2} \frac{rdr}{(\phi(r))^{1/2}} \quad \text{and} \quad \int_{r_0}^{r_2} \frac{Cdr}{r(\phi(r))^{1/2}}$$

diverge, because r_2 is a double root of $\phi(r)$.

If $\phi(r)$ has a double root we have

$$(146) \qquad \phi(r) = (w(r) + h)r^2 - C^2 = 0,$$

$$(147) \qquad \phi'(r) = w'(r)r^2 + 2r(w(r) + h) = 0.$$

If now $\phi(r)$ is negative for values differing slightly from the above then we have a stable circular orbit. If, however $\phi(r) > 0$ for slightly different values of r, then we have instability. This happens if $\phi(r) = 0$ is not a maximum, i.e., if $\phi''(r) \geqq 0$. Then

(148) $w''(r)\, r^2 + 4rw'(r) + 2(w(r) + h) \gtreqqless 0,$

and by means of Equation (147):

(149) $w''(r) + \dfrac{3w'(r)}{r} \gtreqqless 0.$

If this inequality holds, the corresponding circular orbits are unstable; i.e., a small perturbation will make them spiral outward or inward. For more details, see e.g., Lindblad [5], [103] and Coutrez [104].

The instability of the circular orbits has been used by Lindblad in order to explain the spiral arms of the galaxies. He remarked that such orbits can appear only if the oblateness of a spherical system (the quantity $(a - b)/a$ where a is the major axis and b the minor axis) is greater than 0.72. But this is approximately the limit of the oblateness of the elliptic galaxies; systems with greater

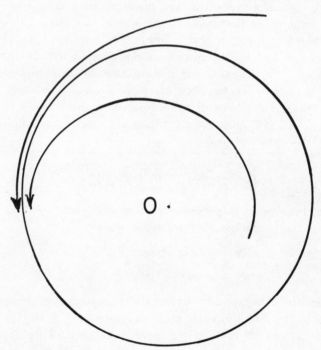

FIGURE 6. Asymptotic Orbits

oblateness usually show spiral arms. However it is believed today that the spiral arms are probably due to other phenomena, and are not asymptotic orbits of the above type.

Orbits of stars in different models of the Galaxy have been calculated by many authors [105], [106], [107], [108], [109], [110] [111], [112], [113], [114], [115], [116], [117], [118], [119]. Such calculations aim at finding the distribution of the pericentra and apocentra of the orbits, their periods, etc.

Another approach to the problem of the plane motions of stars in our Galaxy has been introduced by B. Lindblad [120]. He considers all the motions, that do not differ too much from circular, as epicyclic motions.

If r_1 is the radius of a circular orbit that has the same angular momentum as a given orbit, we introduce the coordinates

$$\xi_1 = r - r_1 \quad \text{and} \quad \eta_1 = r_1\theta,$$

where θ is the angle between the directions of the star and the point moving along the circular orbit.

The equations of motion are

$$(150) \qquad \frac{d^2\xi_1}{dt^2} = \frac{C^2}{(r_1 + \xi_1)^3} - K(r),$$

where $K(r)$ is the force per unit mass at the distance r_1, and

$$(151) \qquad \frac{d\eta_1}{dt} = \frac{Cr_1}{(r_1 + \xi_1)^2} - \omega_1 r_1,$$

where ω_1 is the angular velocity of the circular orbit. Hence

$$(152) \qquad C = \omega_1 r_1^2, \quad \text{and} \quad K(r_1) = C^2/r_1^3 = \omega_1^2 r_1.$$

If now we expand the second members of (150) and (151) in power series of ξ_1 we get to a first approximation

$$(153) \qquad \frac{d^2\xi_1}{dt^2} + \kappa_1^2 \xi_1 = 0,$$

and

$$(154) \qquad \frac{d\eta_1}{dt} = -2\omega_1 \xi_1,$$

where

$$(155) \qquad \kappa_1^2 = 3C^2/r_1^4 + K'(r_1) = 4\omega_1(\omega_1 - A_1),$$

and

$$(156) \qquad A_1 = -\frac{1}{2} r_1 \left(\frac{d\omega}{dr} \right)_1$$

is the first Oort's constant.

The solution is then

$$(157) \qquad \xi_1 = c \cos \kappa_1 (t - t_0)$$

and

$$(158) \qquad \eta_1 = -c \left(\frac{\omega_1}{\omega_1 - A_1} \right)^{1/2} \sin \kappa_1 (t - t_0).$$

These are the equations of an ellipse described by the moving point in its relative motion with respect to a point moving with the circular velocity, and the same angular momentum. If we drop this last restriction and take another circle of reference r_0, where $r_1 = r_0 + c_1$, then we have approximately

$$(159) \qquad \xi = c_1 + c \cos \kappa (t - t_0),$$

$$(160) \qquad \eta = -2Ac_1(t - t_1) - c \left(\frac{\omega}{\omega - A} \right)^{1/2} \sin \kappa (t - t_0)$$

because

$$-2Ac_1 \simeq r_0 c_1 \left(\frac{d\omega}{dr} \right)_0 \simeq r_0(\omega_1 - \omega_0).$$

Here $\kappa^2 = 4\omega(\omega - A)$.

This epicyclic motion can be used in deriving the velocity ellipsoid (see [5]). The velocity components relative to the circular velocity at the point (ξ, η) are approximately

$$(161) \qquad u = \frac{d\xi}{dt} = -c\kappa \sin \kappa (t - t_0),$$

$$(162) \qquad v = \frac{d\eta}{dt} + 2A\xi = 2cB \cos \kappa (t - t_0),$$

where $B = A - \omega$ is Oort's second constant. Hence

$$(163) \qquad c^2 = \frac{u^2}{\kappa^2} + \frac{v^2}{4B^2}.$$

Now if we make the simple assumption that the distribution function F depends only on the parameter c and that the distribution is Gaussian

$$(164) \qquad F = Ae^{-h^2c^2},$$

we get the ellipsoidal law

$$(165) \qquad F = A \exp\left(-\frac{1}{2}\left(\frac{u^2}{a^2} + \frac{v^2}{b^2}\right)\right),$$

and the axes of the ellipsoid are given by

$$(166) \qquad a^2 = \frac{\kappa^2}{2h^2}, \quad b^2 = \frac{2B^2}{h^2}.$$

Then we find

$$(167) \qquad \frac{b^2}{a^2} = 1 - \frac{A}{\omega}.$$

This relation has been derived already in a different way (cf. §III.2). Further work in the same direction has been done by Lindblad and others [5], [121], [122], [123], [124], [125], [126], [127]. Lindblad has considered especially a specific type of epicyclic orbits, the so called dispersion orbits, along which groups of stars disperse because of the differential galactic rotation (see [128], [129], [130], [131], [132]). These orbits have probably some connection with the spiral structure of our Galaxy.

Some important numerical work concerning the motion of a system of rings on the galactic plane has been done lately by P. O. Lindblad [133]. He considers a number of massive points that form one or more rings around the center of the Galaxy (the Galaxy is represented by Schmidt's model [134]). Then he considers the motions of these points under the combined attraction of the Galaxy and of the other points of the rings. It is seen that the rings are preserved for some 10^8 years but they disrupt before 10^9 years in general. Then some formations resembling spiral arms are formed. Such a work is especially important because of its connection with the problem of the formation of the spiral arms.

b. *Three-dimensional orbits.* Calculations of three-dimensional orbits of stars in a galaxy were made by Contopoulos [135], Torgård [136], Ollongren [137], and Hori [138].

If we consider a meridian plane containing the axis of symmetry of the Galaxy (considered as ideally axially symmetrical), which follows a star in its motion around this axis, we get a two-dimensional motion of the star on this meridian plane. In the case that the motion does not extend to infinity, the energy integral defines a torus around the axis of the Galaxy inside which the three-dimensional orbit is contained. The section of this torus with the meridian plane is a closed curve. This curve is called a "curve of zero velocity" because if the moving point reaches this curve its velocity is zero.

The energy integral is

$$(168) \qquad E = \frac{1}{2} \left(R^2 + Z^2 + \frac{C^2}{r^2} \right) + V,$$

where V is the potential energy per unit mass. Therefore the equation of the torus of zero velocity ($R = Z = 0$) is

$$(169) \qquad C^2/2r^2 + V = E,$$

and if

$$(170) \qquad V = \frac{1}{2} \left(-\frac{C^2}{r^2} + P\xi^2 + Qz^2 - \frac{2a\xi^3}{3} - 2b\xi z^2 + \cdots \right)$$

with $\xi = r - r_0$, we have

$$(171) \qquad P\xi^2 + Qz^2 - 2a\xi^3/3 - 2b\xi z^2 - \cdots = 2E.$$

In this case Poincaré's theorem is applicable; i.e., the motions are in general recurrent.

If now the orbits are also ergodic on the energy surface, they fill the whole space inside the curve of zero velocity. However, the existence of a third integral has as a consequence that the orbits fill only part of the available space. This is limited by a curvilinear quadrilateral (see [135]), which in our case is approximately an equilateral trapezium (Figure 7). The orbits are distorted Lissajous figures and the equation of the boundary is given in the form of a series.

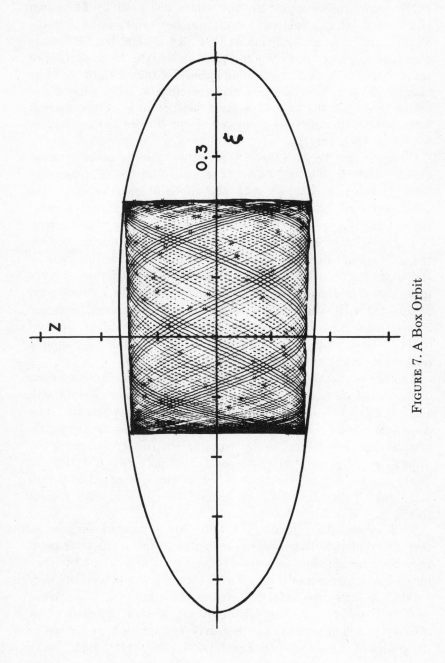

FIGURE 7. A Box Orbit

The same results appear in the orbits discussed by Ollongren and Torård, who calculated a great number of orbits in a model of the Galaxy given by Schmidt [134]. In general the orbits are contained inside a "box" whose apices lie on the torus section of zero velocity. This is a very strong indication that a third isolating integral exists in this case also and therefore in more general potential fields. In the case of a separable potential of the general form (104) the boundary is composed of two ellipses and two hyperbolas (see [84], [138]).

Ollengren and Torgård found some other types of orbits, the so-called "shell" and "tube" orbits. The "shell" orbits fill rings that have no point in common with the curve of zero velocity. The tube orbits fill elongated narrow strips that have, eventually, many folds (Figure 8). These orbits lie near stable periodic orbits.

The calculation of three-dimensional orbits gives the periods of the oscillations along the ξ-axis and the z-axis. In the case of the orbits calculated by Contopoulos the axial oscillations have a mean period of about 24×10^7 years and the z-oscillations a period of about 9×10^7 years (see [84]). Therefore the ξ-oscillations are realized in a time approximately equal to the period of rotation of the Galaxy, while at the same time the orbit makes 2.5 complete oscillations along the z-axis.

Another application of these calculations concerns the dispersion of a group of stars in the Galaxy. If we have a group of stars with slightly differing initial conditions, their velocities when the stars come again near the initial point are very nearly the same, or nearly symmetrical with respect to the axes ξ and z. In fact for any given orbit at every point (ξ, z) there are defined only two directions of motion and 4 velocity vectors, of which two are opposite to the other two. These directions are exactly symmetrical with respect to the axes if $z = 0$.

This means that though the stars are separated because of their differential motions, whenever they return near the same point they have almost the same velocity vectors. Further, even if they have some dispersion in their velocities, their mean periods are very nearly the same (see [184]); therefore the stars of a group come near each other again and again during a time interval of at least some billion years. This fact may explain the persistence of the groups of stars, found by Eggen [139], [140], [141], [142], [143],

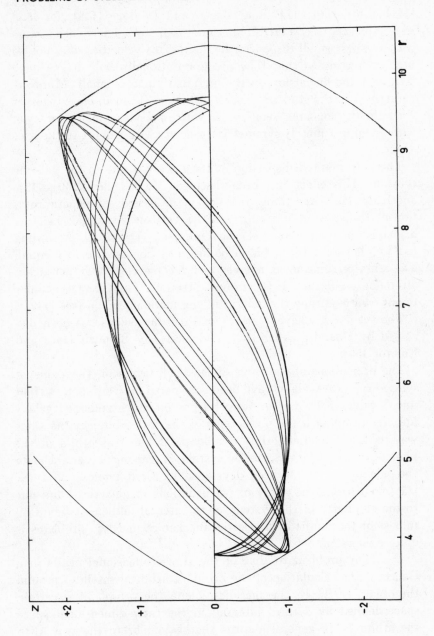

FIGURE 8. One Half of a Tube Orbit (Ollongren)

[144], [145], [146], [147] and Eggen and Sandage [148], for long time intervals (see [149]). These are groups of high velocity stars with a rather small dispersion in their space velocities, but not so near each other so that they are gravitationally bound. We have seen that the dispersion on the meridian plane is small. More accurately after the stars are dispersed they come again close together with almost the same velocity (or with opposite velocity, or even with an approximately symmetrical velocity with respect to the axes ξ and z).

The only marked dispersion of these stars is due to the galactic rotation. This effect has been discussed by Woolley [150], [151] and King [152]. According to Woolley the periods of galactic rotation of the stars of a group must be very nearly equal; i.e., the dispersion of their plane velocities must be very small. According to King, however, the radial velocities may not be so nearly equal. The stars originating at a given point come close together again after one oscillation in the radial direction. This view is similar to our own regarding the motion in the meridian plane (see [149]).

The theory of epicyclic orbits in three dimensions has been discussed by Chandrasekhar [153] and lately by Shimizu [154] and Emoto [155].

The projections of the three-dimensional orbits on the plane of symmetry of the Galaxy have been considered by Contopoulos [156] and Yasuda [157]. The orbits corresponding to a spherical galaxy are osculating with the projections of the real orbits, in the same way as the Keplerian orbits osculate with the real orbits of the planets in the solar system. A system of canonical variables for the galactic orbits has been developed by Contopoulos and Bozis [158]; it permits the study of perturbations of galactic orbits due to the ellipticity of the system, or to external influences. The usefulness of the methods of celestial mechanics in these problems is quite evident.

A similar problem to that of the three-dimensional orbits in a Galaxy is the calculation of the orbits of artificial satellites around the oblate earth. In the meridian plane the orbits fill a certain space defined by a third integral; in the three-dimensional space the orbits fill in general a torus that is similar to the van Allen belts. An application of an approximate third integral of motion to the orbits of satellites was made by Diliberto, Kyner and Freund

[159]. Vinti [160] gave a separable potential that represents approximately the potential of the earth and gives 3 isolating integrals of motion.

c. *Ergodicity*. In the case of galactic orbits, that do not extend to infinity, in 2 or 3 dimensions, one can apply Poincaré's theorem and state that almost all moving points will pass an infinite number of times near the initial point. In fact, in these cases an invariant set Ω of finite measure is always present. In the case of 3-dimensional orbits the section of the above set by the plane of symmetry of the Galaxy is an annulus between two circles of radii r_1 and r_2. Its section with the 3-dimensional space is a torus generated by the curve of zero velocity.

Poincaré's theorem, however, does not tell us if an orbit will fill the whole annulus or the whole torus. This can be decided by means of the so-called ergodic theorem. If the set Ω cannot be separated into invariant parts $\Omega_1, \Omega_2, \cdots$ of positive measure, then it is called metrically indecomposable or metrically transitive. In this case an orbit fills the whole space equally densely. The moving point remains inside any given subset $w \subset \Omega$ of measure m for a fraction of the total time that is asymptotically equal to the ratio m/M, where M is the measure of Ω; i.e., if the moving point stays inside w for time intervals $t_1 + t_2 + t_3 + \cdots = t$ during the total time T, then the limit t/T for $T \to \pm \infty$ exists and is equal to n/M.

This theorem was proved by Birkhoff [161]. Another proof, due to Kolmogorov, is mentioned by Khinchin in his book on the foundation of statistical mechanics [162] ([8]).

The main difficulty of the ergodic theory is to prove that a given system is or is not metrically transitive. No such general theorem is available. Oxtoby and Ulam [165] proved what seems to be the most general theorem in this direction, namely that the measure preserving flows are metrically transitive "in general." However, it is possible that the dynamical systems are exceptional in the sense of Oxtoby and Ulam. (See e.g., Diliberto, [166]). This possibility is closely connected with the existence of some "hidden" integrals of motion. In fact if there exist a number of isolating

---

([8]) For further reading see the monograph of Münster [163] in the Handbuch der Physik. Many mathematical problems of the ergodic theory are discussed by Jacobs [164].

integrals the orbits lie always on the "surfaces" defined by the constant values of these integrals. If, however, only one or two integrals of motion are known, we cannot assert that the orbits will fill the whole space defined by these integrals, because it may be that another isolating or quasi-isolating([9]) integral, not known already, is present. In fact if by calculating some orbits we find that they do not fill the whole permissible space defined by the known integrals, this is an indication of the existence of a further isolating or quasi-isolating integral (see [74], [85]).

In this connection Moser [167], [168] and Arnol'd [169], [170] proved a very important theorem. Under some very general assumptions periodic orbits in a two-dimensional potential field are stable; namely in the 3-dimensional space defined by a constant value of the energy constant the orbits either lie on tori surrounding the periodic orbits, or they are contained between two tori of this kind. In both cases the orbits are not ergodic, i.e., they do not fill the whole energy "surface." In our terminology these cases are either isolating or quasi-isolating.

4. *Some further problems.* The numerical calculation of orbits in different potential fields is of the greatest interest, especially as regards the existence of further integrals.

For example one should calculate some orbits in slightly elliptical systems such as the elliptical clusters, the elliptical galaxies and the oblate earth. In a spherical potential field the orbits fill, in the meridian plane, an area between two circles and two straight lines symmetric with respect to the plane of symmetry (Figure 9). For slightly perturbed fields the area filled by the orbit should be slightly different from the above. This has been found in some examples calculated recently by Contopoulos and Danby. However in some special cases of slightly elliptical systems the orbit fills a parallelogram (see [85]). A number of orbits have been calculated in some resonance cases of the third integral. In the expansion of the third integral in the form of a series there appear an infinite number of divisors of the form $m^2P - n^2Q$ and some of them may be very small or even zero. When one such value is zero we have a resonance in the unperturbed motion (when we have two uncoupled oscillations

---

([9]) A quasi-isolating integral is neither isolating nor ergodic; in general an orbit does not lie exactly on some isolating surface in the phase space, but is contained between two such surfaces (see [74]).

in two perpendicular directions) and the form of the third integral changes (see [152]). It has been found, however, that in the case of higher order resonances, (where $P^{1/2}/Q^{1/2}$ is not equal to 1/1 or to 2/1), the forms of the orbits do not change very much. Only if $P^{1/2}/Q^{1/2}$ is 1/1 or 2/1, do the orbits change appreciably. Such resonance cases should be considered in other problems also.

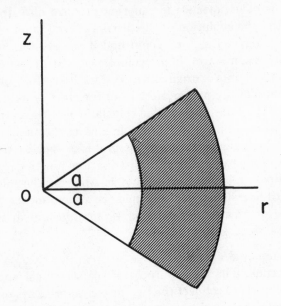

FIGURE 9. Recurrence Region for Spherical Potential

We have calculated some orbits in the case of an axially symmetric system that has not a plane of symmetry. It is evident that a third integral exists in this case too. In the case of a system that has no axial symmetry, there are two integrals besides the energy integral (see [85]). These, however, do not correspond to the angular momentum and the third integral of the axially symmetric case.

One could think of other more complicated potential fields also. Such is the case of the 3-body problem or the n-body problem. In all these cases the calculation of some orbits may indicate the existence of new isolating or quasi-isolating integrals, whenever such integrals exist.

Of special importance is the calculation of periodic orbits. The periodic orbits play an important role in classical celestial mechanics and they should play a similar role in stellar dynamics. A few periodic orbits in the case of the potential field (131) have been calculated by Goudas and Barbanis [171]. The initial velocities of these orbits are perpendicular to the plane of symmetry. We found also periodic orbits with nonvertical initial velocities.

The calculation of orbits in a realistic potential field, representing our Galaxy is of importance for another problem also. In this way one could find the evolution of the distribution of stars of different ages in the Galaxy. E.g., one could find if the A stars were more concentrated in spiral arms in previous times, and so find eventually the age and the points of origin of these stars. This work necessitates accurate values of the proper motions of the stars as well as a good knowledge of the galactic field. The calculations of the perturbations due to the spiral arms is of interest in this connection.

Another application of the third integral is in constructing models of stellar systems by means of a distribution function F that depends on three arguments, E, C, and Φ. Most of the models constructed until now depend only on one or two arguments. Therefore it should be of great interest to have more general distributions, including the third integral also.

IV. The statistical approach.

1. *Introduction*. The main purpose of the statistical approach of stellar dynamics is to connect the two former approaches of particles and continuum. This is essentially the object of statistical mechanics in general. In fact in statistical mechanics a probability is defined for any given particle to be in any given region of the phase space. The distribution of a great number of particles is assumed to follow this probability law which gives a continuum distribution.

This method, however, is not easily applicable to stellar dynamics; in fact a consistent definition of probability in the case of stellar dynamics has not yet been given.

A "probability density" p is a positive quantity that is normalized, i.e., $\int_\mu p\,dw = 1$ for all the phase space μ, and it is an integral of the equations of motion. The last property is derivable from the following fact. If the probability that a particle belongs to a certain set w in the phase space is P, then after some time t the set w is transformed into $\phi(w,t)$ and this has the same measure as w. On the other hand the probability of $\phi(w,t)$ is equal to P, because each point of w defines uniquely one point of $\phi(w,t)$. Therefore the probability density $p = \lim(P/m(w))$ is constant along an orbit.

We can define as probability density the quantity $p = f/N$, where f is the distribution function and N the total number of stars of a stellar system. Such a probability density would give only the known continuum approach of stellar dynamics. However, here we want to *find f*, as it is done, for example, in classical statistical mechanics, by using the ergodic theorem or assuming equal a priori probabilities.

The form of f may be found by adding a new element in our picture, and this is the interaction of the stars among themselves. In the continuum approach we never used the fact that the individual stars which compose the continuum attract each other according to Newton's law. If these interactions of the stars with each other are taken into account, we may be able to derive a "final" distribution function towards which any initial distribution of positions and velocities tends, or at least to find the evolution of a distribution function in the course of time, once the initial distribution function is given.

This problem can be solved, in principle, by numerical integration. This, however, is equivalent to solving completely the n-body problem, and this is quite impossible in practice since the number of the gravitating bodies is very great. Therefore, we should rely on some general principles.

The basis of statistical mechanics is the assumption of equal a priori probabilities for equal volumes of the phase space. In order to do the same in stellar dynamics it is sufficient to find an invariant set Ω of positive but finite measure; then a function p that is a positive constant inside Ω and zero outside it can be taken as a probability density. Kurth [4], [172], however, remarks that no such invariant set is known and probably does not exist; therefore no statistical mechanics of stellar systems exists. For this reason he rejects all statistical methods in dynamical astronomy, such as the notion of the relaxation time, the Boltzmann distribution etc.

However there is at least one case, mentioned above (in §II.4) where we can define an invariant set Ω. This is the case of the plane restricted 3-body problem. In this case a probability density can be defined. It should be very important to find similar cases in the n-body problem also. The results of Arnol'd [33] open new possibilities in this direction.

Further one may find criteria to distinguish between orbits extending to infinity and recurrent orbits. Then we might apply our considerations to the set of the recurrent points only.

But even if these possibilities are not realized, one can apply quite generally the methods of statistical mechanics to stellar dynamics, under some "reasonable" assumptions, as a convenient means to study approximately the behavior of stellar systems. Such a discussion has been made by Heckmann [173].

The statistical mechanics of stellar systems can be applied in a region of the phase space that is not infinite, whether this region is really invariant or not, if we have reasons to assume that the motions that extend outside this region are rather exceptional. This method is of course, only approximate, but it has the same value as the approximation of a stellar system by a continuum, or even by n gravitating bodies moving according to Newton's law.

In fact the main argument of Kurth against the statistical mechanics of stellar systems is that it contradicts the theorems of Hopf. The statistical mechanics considers the evolution of a stellar system as an irreversible process, i.e., as a one-way evolution toward a final state which we might call the state of maximum entropy. According to the theorems of Hopf, however, there is no such final state. The system is either recurrent or its representative point goes to infinity; but the second alternative happens only if the representative point came from infinity. We may say more loosely that a system disintegrates only if it was formed by captures. A recurrent system, however, is almost periodic for infinite time, and no final state exists, unless it is periodic.

This objection is quite similar to the famous "recurrence objection" (Wiederkehreinwand) of E. Zermelo against the irreversibility of statistical mechanics. Zermelo's argument, based on Poincaré's theorem, states that the increase of entropy of a system is followed by a corresponding decrease of entropy as the system returns arbitrarily near its initial state. In fact, such increases and decreases of entropy alternate, and evolution of every physical system is oscillatory.

A very interesting "conciliation" of the recurrence theorem with the classical views of thermodynamics is due to P. and T.

Ehrenfest [174]. The Ehrenfests point out (a) that if at some time the entropy is much less than its maximum value, it will practically always tend to increase, (b) that this is true for both time directions, and (c) that the entropy remains quite near its maximum value practically all the time. In other words the time interval during which a system is far from equilibrium is quite small in comparison with the time of equilibrium. Therefore, the actual Universe is a very improbable deviation from equilibrium.

However, such deviations from equilibrium do exist, although they are very rare. In the case of a gas the time of relaxation, i.e., the time needed to reach equilibrium, is of the order of 10^{-9} seconds; whereas the recurrence time for one cm^3 of gas (10^{18} atoms), i.e., the time needed so that each atom should return near its initial position with nearly the same velocity, has been calculated by Boltzmann to be of the order of $10^{10^{19}}$ years! In the case of a stellar system the relaxation time is of the order of 10^7 to 10^{14} years, while the recurrence time is at least of the order of $10^{1,000}$ to $10^{1,000,000}$ years as we have already calculated (in §II.5). Therefore, even if a stellar system is recurrent, the statistical methods can be applied if they deal with time intervals small in comparison with the recurrence time.

The Ehrenfests' approach has many advantages, but it has one main disadvantage in that it considers the two time-directions as essentially equivalent. However, the uniqueness of the time direction is one of the most basic principles of physics, of the same status as the principle of causation.

On the other hand, the gravitational law of Newton and the laws of mechanics in general are symmetric with respect to the past and future. One can escape from the dilemma only by assuming a limited extrapolation of any physical law and the laws of mechanics in particular. One can practically never extrapolate the laws of mechanics for intervals as long as the recurrence times. For example, the accuracy with which the changes of the angular elements of the planets are known, on which the accuracy of the law of Newton is based, is of the order of $1''$/century (see [175]). This corresponds to an inaccuracy of $90°$ in about 4×10^{12} years, which is roughly of the order of one relaxation time but quite small in comparison with the recurrence time.

If we assume that Newton's gravitational law changes slowly in time then our proof of Hopf's theorem is no more valid. In practice there are many quite evident reasons why one cannot extrapolate Newton's law for extremely long time intervals. The loss of mass of the stars, their friction with the interstellar matter, or the actual collisions of stars, completely change the behavior of a stellar system in time intervals quite small in comparison with the recurrence time. Fricke [176] has calculated the effects due to a resisting medium on the motion of the stars. From his formulae we find that for a density of the interstellar matter of the order of 10^{-24} gram^{-3}, the velocity of a star is reduced to one half after a time of the order of a hundred to a thousand times of relaxation. Therefore, Newton's law is applicable only for time intervals of the order of a thousand times of relaxation. In much longer times the small irreversible effects become predominant. Hence in the long run all the physical and astronomical phenomena show a definite time-direction.

The theorems of Hopf are very useful as regards the mathematical theory of the n-body problem, but not as regards the behavior of real stellar systems. For these reasons the use of statistical methods in stellar dynamics is well justified; in fact in many cases these methods give us more information than particle mechanics.

2. *The Relaxation Time.* As a first illustration of the application of statistical methods to the problems of stellar dynamics we shall give the main steps of the calculation of the time of relaxation of a stellar system. The first satisfactory calculation of the time of relaxation was made by Chandrasekhar [2].

There are several definitions of the time of relaxation. Here we shall use the following: The time of relaxation T_E is the time needed so that the mean change of the kinetic energy of a star due to encounters is equal to its initial kinetic energy. Namely, if E is the kinetic energy of a star and ΔE is the energy exchange in each encounter, then in time T_E we get the equality

$$(172) \qquad\qquad \sum (\Delta E)^2 = E^2.$$

This time is found in the following way: Each encounter is considered approximately as a two-body problem and $(\Delta E)^2$ is calculated; then we take the contribution of all the encounters suffered by a star in time Δt. The sum $\sum (\Delta E)^2$ becomes equal to E^2 when $\Delta t = T_E$.

Let $\overline{v_g}$ be the velocity of the center of mass of two stars with masses m_1, m_2 and velocities $\overline{v_1}, \overline{v_2}$; i.e.,

$$(173) \qquad \overline{v_g} = \frac{1}{m_1 + m_2} \, (m_1\overline{v_1} + m_2\overline{v_2}).$$

On the orbital plane (where the relative motion of the two stars takes place) we have the relative velocity

$$(174) \qquad \overline{v} = \overline{v_2} - \overline{v_1}.$$

Hence

$$(175) \qquad \overline{v_2} = \frac{m_1\overline{v}}{m_1 + m_2} + \overline{v_g}.$$

The relative velocity does not change in measure during the encounter, but it is deflected by an angle $\pi - 2\psi$ (Figure 10). D is

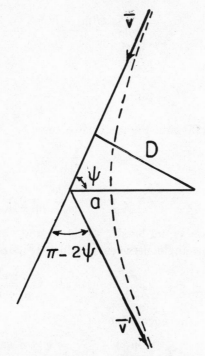

FIGURE 10. Change in Relative Velocity

the distance of m_2 from the initial velocity vector of m_1 and is called the "impact parameter". If

$$(176) \qquad \frac{x^2}{a^2} - \frac{y^2}{b^2} = 1$$

is the equation of the relative orbit, then

$$(177) \qquad \tan^2 \psi = b^2/a^2,$$

and

$$b = D,$$

while

$$(178) \qquad b^2/a = p = C^2/(G(m_1 + m_2)),$$

where

$$(179) \qquad C = \text{The constant of areas} = Dv.$$

Therefore

$$(180) \qquad a = \frac{G(m_1 + m_2)}{v^2};$$

hence

$$(181) \qquad \tan^2 \psi = \frac{D^2 v^4}{G^2 (m_1 + m_2)^2}.$$

Then (see (175) and Figure 11) we have

$$\bar{v}_{2g} = \bar{v}_g - \bar{v}_2 = \frac{m_1}{m_1 + m_2} \bar{v},$$

$$\bar{v}'_{2g} = \bar{v}_g - \bar{v}'_2 = \frac{m_1}{m_1 + m_2} \bar{v}'.$$

As $\bar{v}$ and $\bar{v}'$ have equal length, so do $\bar{v}_{2g}$ and $\bar{v}'_{2g}$.

Let Φ be the angle between $\bar{v}_g$ and $\bar{v}$ (Figure 11); then

$$(182) \qquad v_2^2 = v_g^2 + \frac{2m_1}{m_1 + m_2} v_g v \cos\Phi + \left(\frac{m_1}{m_1 + m_2}\right)^2 v^2.$$

Similarly after the encounter

$$(183) \qquad v_2'^2 = v_g^2 + \frac{2m_1}{m_1 + m_2} v_g v \cos\Phi' + \left(\frac{m_1}{m_1 + m_2}\right)^2 v^2,$$

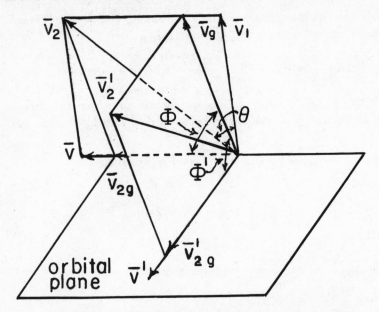

FIGURE 11. Three-Dimensional Aspect of Encounter

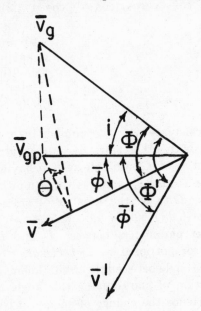

FIGURE 12. Projection Angles

because the length of v remains the same. Hence the change in kinetic energy during the encounter is

$$(184) \qquad \Delta E = \frac{1}{2} m_2(v_2'^2 - v_2^2) = \frac{m_1 m_2}{m_1 + m_2} v_g v(\cos \Phi' - \cos \Phi).$$

If i is the angle between $\bar{v}_{gp}$, the projection of $\bar{v}_g$ on the orbital plane, and $\bar{v}_g$ and $\bar{\phi}$, $\bar{\phi}'$ the angles between $\bar{v}$, $\bar{v}'$ and $\bar{v}_{gp}$ (Figure 12), then

$$(185) \qquad \cos \Phi = \cos \bar{\phi} \cos i, \qquad \cos \Phi' = \cos \bar{\phi}' \cos i.$$

But

$$(186) \qquad \bar{\phi}' - \bar{\phi} = \pi - 2\psi.$$

Hence

$$(187) \qquad \frac{\bar{\phi} + \bar{\phi}'}{2} = \frac{\pi}{2} - \psi + \bar{\phi},$$

therefore

$$\cos \Phi' - \cos \Phi = \cos i(\cos \bar{\phi}' - \cos \bar{\phi}) = 2 \sin \frac{\bar{\phi} + \bar{\phi}'}{2} \sin \frac{\bar{\phi} - \bar{\phi}'}{2} \cos i$$
$$(188)$$
$$= -2 \cos i \cos(\bar{\phi} - \psi) \cos \psi$$

and

$$(189) \qquad \Delta E = -\frac{2m_1 m_2}{m_1 + m_2} v_g v \cos(\bar{\phi} - \psi) \cos \psi \cos i.$$

Let us calculate now the number of encounters of a star with velocity v_2 with other stars. If (v_1, θ, ϕ) are the polar coordinates of the point $\bar{v}_1$ in a coordinate system having as z-axis the velocity $\bar{v}_2$ (θ is the angle $\bar{v}_1\bar{v}_2$, and ϕ is the azimuth of the plane $\bar{v}_1\bar{v}_2$). Let $N(v_1, \theta, \phi) \, dv_1 d\theta d\phi$ be the number of stars with velocities in the range $(v_1, v_1 + dv_1)$, $(\theta, \theta + d\theta)$, $(\phi, \phi + d\phi)$. The number of stars with impact parameters between D and $D + dD$ encountered in time dt is proportional to $2\pi D d D D v \, dt$.

If Θ is the angle between the orbital plane and the plane $\bar{v}_1\bar{v}_2$, then the proportion of stars having this angle between Θ and $\Theta + d\Theta$ is $d\Theta/2\pi$. Hence the change of energy of v_2 in time dt, due to encounters with stars with the parameters v_1, θ, ϕ, D, Θ in the above range, is

$$\sum (\Delta E)^2 (v_1, \theta, \phi, D, \Theta)$$

(190)

$$= (\Delta E)^2 2\pi N(v_1, \theta, \phi)\, vD\, dD \frac{d\Theta}{2\pi}\, dv_1\, d\theta\, d\phi\, dt$$

$$= 4N(v_1, \theta, \phi) v_g^2 v^3 \left(\frac{m_1 m_2}{m_1 + m_2}\right)^2 \cos^2 i \cos^2(\overline{\phi} - \psi) \cos^2\psi\, DdD d\Theta\, dv_1\, d\theta\, d\phi\, dt.$$

Now we must integrate over all the values of D, Θ, θ, ϕ and v_1 to find the total change $\sum (\Delta E)^2$ in time dt. By (181)

(191)
$$DdD = \frac{G^2(m_1 + m_2)^2}{v^4} \frac{\sin \psi}{\cos^3\psi} d\psi,$$

and for $D \to 0$ we have $\psi \to 0$ (because the deviation $\pi - 2\psi$ then tends to 2π), while for $D \to \infty$, $\psi \to \pi/2$ (the deviation tends to zero).
Hence

$$\sum (\Delta E)^2 (v_1, \theta, \phi, \Theta)$$

(192)

$$= 4N(v_1, \theta, \phi)\, G^2 m_1^2 m_2^2 \frac{v_g^2}{v} \cos^2 i \int_0^{\psi_0} \cos^2(\overline{\phi} - \psi) \tan \psi\, d\psi\, d\Theta\, dv_1\, d\theta\, d\phi\, dt.$$

This becomes infinite logarithmically as $\psi_0 \to \pi/2$. However when D is great we cannot consider the encounters as two-body problems. If D_0 is the distance of the neighboring stars, and $D \gg D_0$ there are many encounters at the same time which probably cancel each other. On the other hand if $D < D_0$ the formula (192) over-estimates each $(\Delta E)^2$ because an actual encounter lasts less time than in the case of a two-body problem. For this reason Chandrasekhar takes $D = D_0$ as the upper limit of the impact parameter. It is to be noted that a change of this limit by a factor of 2 does not appreciably change the value of $(\Delta E)^2(v_1, \theta, \phi, \Theta)$ because it contains the term

$$\int_0^{\psi_0} \cos^2(\overline{\phi} - \psi)\, d \log \cos \psi,$$

and this does not change very much if $\tan \psi_0$ changes by a factor of 2.

This is the most delicate point in the calculation of $\sum (\Delta E)^2$. The integrations with respect to Θ, θ, ϕ are made relatively easily if the distribution of velocities is assumed to be spherical (i.e., $N(v_1, \theta, \phi) = N(v_1) \sin\theta/4\pi$), and if we drop all the numerically unimportant terms. Finally the integration with respect to v_1 is effected by assuming a Maxwellian distribution of the velocities v_1, i.e.,

$$(193) \qquad N(v_1)\, dv_1 = \frac{4j^3}{\pi^{1/2}}\, N \exp(-\, j^2 v_1^2)\, v_1^2 dv_1,$$

where N is the number of stars per unit volume.

Finally we find

$$(194) \qquad \frac{\sum (\Delta E)^2}{E^2} = \frac{dt}{T_E}$$

where

$$(195) \qquad T_E = \frac{v_2^3}{32\pi\, NG^2 m_1^2 G(x_0) \ln\left(\dfrac{D_0 v_2^2}{G(m_1 + m_2)}\right)}.$$

Here

$$(196) \qquad x_0 = jv_2 \quad \text{and} \quad G(x_0) = \frac{1}{2x_0^2}\, (H(x_0) - x_0 H'(x_0)),$$

where H is the error function

$$(197) \qquad H(x_0) = \frac{2}{\pi^{1/2}} \int_0^{x_0} \exp(-\, x^2)\, dx.$$

Hence if $dt = T_E$ then $\sum (\Delta E)^2 = E^2$; therefore T_E is the time of relaxation.

For our Galaxy $N = 0.1\,\text{star/pc}^3$, $D_0 = 2.7\,\text{pc}$, $m_1 = 0.5\odot$, $v_2 = 20\,\text{km/sec}$, $j = 0.032\,\text{sec/km}$ and we find that

$$T_E = 7 \times 10^{13}\ \text{years}.$$

This quantity is big enough so that the individual orbits of stars do not change appreciably because of the encounters. During the revolution of the sun around the center of the Galaxy (2×10^8 years) the mean change of the energy of a star is about $0.0017E$. Correspondingly the mean change in the direction of the motion of a star is $\Delta\psi \cong 0°.07$. For this reason we have neglected the stellar encounters in calculating the orbits of stars in a stellar system.

If the velocities v_2 also have a Maxwellian distribution we find a mean time of relaxation

(198)
$$\overline{T_E} = \frac{1}{16} \left(\frac{3}{\pi}\right)^{1/2} \frac{(\overline{v_2^2})^{3/2}}{NG^2m^2\ln(D_0\overline{v_2^2}/2Gm)},$$

if the masses of all the stars are equal.

In the case of a spherical cluster we may apply the virial theorem and have

(199)
$$\overline{v^2} = \frac{Gmn}{2R^*}$$

where n is the number of stars, and $R^* = r^*/2$ is the value used by Chandrasekhar as the radius of the cluster. If we draw around each star a sphere of radius $D_0/2$, the volume of all these spheres is approximately equal to the volume of the cluster; i.e.,

(200)
$$\frac{4}{3}\pi\left(\frac{D_0}{2}\right)^3 n = \frac{4}{3}\pi R^{*3},$$

hence

(201)
$$\frac{D_0}{R^*} = 2n^{-1/3}.$$

Also

(202)
$$\frac{4}{3}\pi R^{*3}N = n.$$

Setting these values in formula (198) we get

(203)
$$\overline{T_E} = \frac{1}{16}\left(\frac{3\pi}{2}\right)^{1/2}\left(\frac{nR^{*3}}{Gm}\right)^{1/2}\frac{1}{\ln(n/2^{3/2})}.$$

As an example, the time of relaxation of the Pleiades is $\overline{T_E} = 2.9 \times 10^i$ years. Therefore in this case the time of relaxation is much shorter than the age of the cluster; i.e., the encounters must have already achieved a statistical equilibrium. The same thing happens in the central parts of the globular clusters also (see [177], [178]).

Many astronomers like Charlier, K. Schwarzschild, Rosseland, Heckmann and Siedentopf, Spitzer, and Fricke have worked also on the problem of the time of relaxation. Fricke [179] has pointed out some errors in the earlier discussions. However, Fricke's method also (which is presented in von der Pahlen's book [3]) is not convenient, as pointed out by Heckmann [180].

Chandrasekhar has made another approach to the same problem using purely statistical methods (see [181]). He finds almost exactly the same values for the time of relaxation.

The above calculations, however, consider only the interactions of stars among themselves. Some authors have considered also the effect of the interstellar medium on the relaxation of stellar velocities. Spitzer and Schwarzschild [182], [183] have studied the effect of the interstellar clouds on the motions of the stars. They found that because of the encounters of the stars with interstellar clouds the time of relaxation is shortened very much and is only some 10^8 years for our Galaxy. This is due to the fact that $\overline{T}_E$ is inversely proportional to $m^2 N$, where mN is the mass density. Hence if the mass density remains the same, but instead of stars of mean mass $0.5\odot$ we have big clouds of mass $m = 10^5\odot$, the time of relaxation is 2×10^5 times smaller. Only for the high velocity stars is the time T_E still big enough, since it is proportional to $v_2^{3/2}$.

Osterbrock [184] found similar numerical results by assuming that the stars move in a fluctuating interstellar medium. Similar conclusions are also reached by Ogorodnikov [185].

A few years ago Hénon [186] calculated the time of relaxation in a way different from that of Chandrasekhar. He considered the distant encounters as perturbations instead of two-body encounters. This is more reasonable than completely disregarding the distant encounters.

Hénon's formula is similar to that of Chandrasekhar, except that the logarithmic term of Equation (203) is replaced by

$$\ln(2\overline{v}^3 T/e^2 G(m_1 + m_2)).$$

This means that the perturbations $\sum (\Delta E)^2$ are not proportional to T, but increase initially a little faster than T. This, however, seems to be due to the method of calculation used and not to a real physical phenomenon.

When $\overline{v}T$ becomes $\geq \lambda$, where λ is the "Jeans length" (the diameter of the condensations formed by gravitational instability (see [69, p. 348]), then $\overline{v}T$ is replaced by

$$(204) \qquad \lambda = \left(\frac{\pi v^2}{3GNm} \right)^{1/2} = \frac{2\pi r^*}{3} \cong 2r^*;$$

r^* is the radius of the cluster given in §II.1 by the formula $r^* = GM/v^2$, and $Nm = 3M/(4\pi r^{*3})$. Therefore the "Jeans length" is roughly equal to the diameter of a cluster. Hénon gives D_0 by the formula

$$(205) \qquad D_0 = N^{-1/3} = \left(\frac{3n}{4\pi r^{*3}} \right)^{-1/3} = \left(\frac{4\pi}{3} \right)^{1/3} r^* n^{-1/3} \ll 2r^*,$$

hence

$$\lambda \gg D_0.$$

The logarithmic term in Equation (203) becomes

$$\frac{3}{2} \ln \left(\frac{2\pi n}{3e^2} \right) ;$$

i.e., Chandrasekhar's time of relaxation $\overline{T}_E$ must be multiplied by about $2/3$. The change is unimportant.

The same formulas apply also in the case of a plasma (see [187], [188]). In this case

$$(206) \qquad \lambda = \left(\frac{kT}{8\pi Ne^2} \right)^{1/2}$$

(T is absolute temperature, k is the Boltzmann's constant, N the particle density, and e the electron's charge), and this is the so-called "Debye length," i.e., the distance over which the Coulomb forces are effective in a plasma. At greater distances a screening effect due to charges of opposite sign does not permit the interaction of the charges.

In the case of a stellar system this screening effect does not appear, and therefore λ is equal to the diameter of the system (see [189]). However it should be of interest to find out the action of distant perturbations, in connection with the problem of the fragmentation of a system into clusters of stars. Jeans' study refers to instabilities that occur in a continuum if the wavelength of a density variation exceeds a critical value which is the Jeans length. Similar work was done lately by Lynden-Bell [190]. The problem now is to find how far these considerations apply to real stellar systems composed of interacting stars and not only to continuum media. Layzer [191] has recently criticized Jeans' results; he points out that nonlinear effects probably do not permit fragmentation.

In this connection one should consider the problem of formation of condensations of different sizes (galaxies, stellar clusters, stars) especially in rotating systems. It is not quite clear if formula (204) can be applied also when $\bar{v}$ refers to a rotating coordinate system, i.e., if it is the velocity with respect to the local standard of rest.

Some experimental calculations of the time of relaxation in a plasma have given values that are 1000 times smaller than calculated theoretically. This is the so-called Langmuir's paradox (see [192], [193]). Fricke [66] has suggested that perhaps forces other than Coulomb's play a role here. A few years ago, however, Gabor, Ash and Dracott [194] showed that some oscillatory fields are generated at the walls of the plasma containers adding energy and entropy to the electrons, and thus causing a Maxwellian distribution to be reached very soon.

A problem similar to that of the time of relaxation is the problem of dynamical friction (see [195], [196], [197], [198], [186]), i.e., the continuous decrease of the velocity of a star along its orbit due to interactions with other stars.

Chandrasekhar's time of relaxation refers to systems with constant density. In actual systems, however, the density decreases outwards. Therefore the time of relaxation in the outer parts of a stellar system is larger. Consequently a system cannot be very approximately in equilibrium except in its central parts. Further, the time of relaxation for a circular orbit is larger than for a radial orbit that has the same apocenter. Calculations of the time of relaxation in different models of star clusters were made by Woolley and Robertson [200], [71] and by King [201]. King finds a mean time of relaxation 4 times greater than given by Chandrasekhar's formula, for clusters with a central condensation.

3. *The Escape of Stars from a Cluster.* One of the most important problems in the evolution of a stellar cluster is the escape of stars from it and the subsequent dissolution of the cluster. This problem has been studied by a number of authors, including Chandrasekhar, King, Hénon, von Hörner, etc. In order that a star should escape from a cluster its total energy must be positive or zero; i.e., its velocity must become equal or greater than v_∞, where

$$(207) \qquad \frac{1}{2} m_i v_\infty^2 = G \sum_{j=1; i \neq j}^{n} \frac{m_i m_j}{r_{ij}} \qquad (i \neq j),$$

and further encounters should not reduce the velocity at any point below the corresponding v_∞. We may omit the latter effect as of unlikely occurrence in general (see [199]).

Adding all similar terms we find

$$(208) \qquad \frac{1}{2} \sum_{i=1}^{n} m_i v_\infty^2 = \frac{1}{2} \, mn\overline{v_\infty^2} = -2v = \frac{Gm^2 n}{R^*} \quad \text{or} \quad \overline{v_\infty^2} = \frac{2Gmn}{R^*}.$$

On the other hand from the virial theorem we have $\overline{v^2} = Gmn/2R^*$, hence

$$(209) \qquad \overline{v_\infty^2} = \overline{4v^2}$$

In a Maxwellian distribution of velocities

$$(210) \qquad dn = \frac{4nj^3}{\pi^{1/2}} \exp(-j^2 v^2) \, v^2 dv,$$

we have

$$(211) \qquad \overline{v^2} = 3/2j^2, \quad \text{and} \quad \overline{v_\infty^2} = 6/j^2.$$

Hence the fraction of stars

$$(212) \qquad Q = \frac{4}{\pi^{1/2}} \int_{6^{1/2}}^{\infty} \exp(-x^2) \, x^2 dx = 0.0074$$

have a velocity greater than the velocity of escape.

As the stars escape, however, new stars acquire velocities greater than the velocity of escape because of the encounters, and so we have a continuous escape of stars. It is argued that during one time of relaxation, T_E, a Maxwellian distribution is restored in a cluster and therefore in n_1 times of relaxation a fraction of $\Delta n/n = n_1 Q$ stars will escape. Hence in general the proportion of stars escaping in time Δt is

$$(213) \qquad \frac{\Delta n}{n} = 0.0074 \frac{\Delta t}{T_E}.$$

In other papers Chandrasekhar [195], [196], [197] calculates the probability $Q(\tau)$ that a star will escape from a cluster in time τ. He finds first that if we omit the action of dynamical friction, the cluster is dissolved completely in a few times of relaxation. An open cluster is dissolved in a few 10^8 years. If, however, dynamical friction is taken into account, the lifetime of the clusters is much longer. Chandrasekhar finds then that $Q(\tau) = 1 - \exp(-\tau/T_0)$ where T_0 is the half-life of a cluster equal to $T_0 = 266\,\overline{T}_E\,(^{10})$. Hence the half-life of the Pleiades instead of being of the order of 3×10^7 years becomes of the order of 3×10^9 years.

Recently the problem of the ejection of stars from a cluster has been considered in detail by many authors.

King in a series of papers [199], [201], [202], [203], [204], made some improvements on Chandrasekhar's theory. He considered the variation of the time of relaxation $\overline{T}_E$ with time. As the stars escape, the remaining cluster contracts continuously and $\overline{T}_E$ decreases. Therefore the total life-time of a cluster is found to be only about $40\overline{T}_{E0}$, where $\overline{T}_{E0}$ is the initial time of relaxation.

Similar results were found by von Hörner [205]. He finds a dissolution time that is about $30\overline{T}_{E0}$. The difference between King and von Hörner is that King assumes that escaping stars take almost no energy with them, whereas von Hörner assumes that they take a fraction of the total energy equal to 0.0052 (equal to the fractional energy of the escaping stars in a Maxwellian distribution, when they reach an infinite distance from the center) during one time of relaxation. Earlier L. Spitzer [206] supposed that each star takes away an amount of energy $m\overline{v^2}/2$ where $\overline{v^2}$ is the mean square velocity. This is, of course, an overestimate. Probably King is nearer to the truth, because a star escapes just as it acquires the escape velocity and probably it takes away only an insignificant amount of energy. It is evident that the Maxwellian distribution cannot be valid for velocities bigger than the escape velocity.

After most of the stars of cluster escape there will remain probably only a stable multiple star, or even a double star. King found that if this double star gets all the energy of the cluster it will be just a normal double star.

Hénon [41], [42] has made a critical study of the escape mechanisms of a star from a cluster. He points out that a gradual increase of stellar velocities cannot result in the ejection of stars; because if a star gets a velocity near the velocity of escape, it moves in a very elongated orbit and only rarely comes near the center where encounters take place. In fact, the star never acquires

---

(^{10}) The formula used by Chandrasekhar is $T_0 = 133\eta_0^{-1}$, where $\eta_0^{-1} = 2T_E$.

the escape velocity if it is only subjected to a continuous set of very small perturbations. Only during a very close encounter can a star acquire abruptly the escape velocity. Such encounters, however, are very rare.

Hénon finds that the escape rate is given by the formula

$$(214) \qquad \frac{dn}{dt} = -0.00426 \left(\frac{Gmn}{r_0^3} \right)^{1/2}$$

for the Plummer model, where n is the number of stars and r_0 is the radius including, in projection, one half of the total mass. In the case of a globular cluster ($n = 10^5$) this quantity is about 150 times smaller than the value

$$(215) \qquad \frac{dn}{dt} = -0.125 \, (\log_{10} n - 0.45) \frac{Gmn}{R^{*3}}$$

derived from Chandrasekhar's formulas, as given in [2] ([11]). It is also about 40 times smaller than the corresponding value given by King. Similar calculations were made recently by Woolley and Dickens [207].

Another important factor that influences the rate of escape of stars from a cluster is the effect of the tidal field of the Galaxy and of the passing-by stars and interstellar clouds. This problem also has been considered by a number of authors.

The effect of the tidal force of the Galaxy has been considered by Bok [208], Mineur [209], Chandrasekhar [2], von Hörner [205], King [210] and others. In order that a cluster should not be unstable under the tidal forces of the Galaxy, its density must be higher than a certain limit, which is about $0.1 \odot / \mathrm{pc}^3$. This condition is realized in the real clusters; e.g., in the Pleiades the density is about 15 times the critical density. However, if a star goes out farther than a limiting distance from the center of the cluster it is usually lost from it because of the tidal forces. The limit according to von Hörner is

$$(216) \qquad r_s = R_0 \left(\frac{M}{2M_g} \right)^{1/3},$$

where M is the mass of the cluster, M_g the mass of the Galaxy

---

([11]) If we insert in (215) the value of G in units of parsec, solar mass and year, we get the formula (5.314) of Chandrasekhar.

and R_0 the distance from the center of the Galaxy. If we set M_g = $2 \times 10^{11} \odot$ and $R_0 = 10$ kpc, we get

$$(217) \qquad r_s = 13.6 \left(\frac{M}{1000} \right)^{1/3},$$

where r_s is given in parsecs and M in solar masses.

If the radius of a cluster approaches this limit, its dissolution is very fast. For $R^* = 0.5 r_s$ von Hörner finds a time of dissolution equal to $8 \overline{T}_{E0}$. If R^* is smaller the life time increases up to the limit $30 \overline{T}_{E0}$.

The effect of the encounters of a cluster with other stars on the rate of ejection of stars is small. However it has been found (Spitzer [211]) that clouds of interstellar gas going by a cluster exert a much greater influence than passing-by stars, namely about 30 times greater. This effect is of the same order of magnitude as the tidal effect of the whole Galaxy on a cluster. Thus the disruption times of the actual clusters are of the order of 10^8 to 10^9 years.

King [203] has considered both the contraction of an open cluster due to the ejection of its stars and its gradual expansion due to the effect of the tidal forces. The loose clusters tend to be dissolved by tidal forces, while the dense clusters are dissolved mainly by internal encounters. King finds a critical radius r_{crit} (about $1\frac{1}{2}$ pc for a cluster of 100 stars), that separates the clusters into 2 classes. The clusters with $r < r_{\text{crit}}$ are contracting, while those with $r > r_{\text{crit}}$ are expanding as their energy increases due to encounters with clouds. In both cases the maximum lifetime of an open cluster is of the order of 10^9 years.

In a recent paper King [210] has considered three factors that define the structure of a cluster; namely the number of its stars, its energy, and the effect of the galactic tidal field. These correspond to the three parameters that enter the empirical density law of a cluster given by King

$$(218) \qquad \rho = k \left[\frac{1}{(1 + (r/r_c)^2)^{1/2}} - \frac{1}{(1 + (r_t/r_c)^2)^{1/2}} \right]^2.$$

In this formula r_t gives the cutoff due to the galactic tidal field, r_c gives the concentration of the cluster, and k is proportional to the number of stars. Formula (218) was shown by King to represent remarkably well the real density distribution of the globular clusters and probably of the galactic clusters and dwarf elliptical galaxies also. The encounters do not seem to change the general structure of the clusters; they only slowly change the parameters entering in the above formula. King ascribes the similarity of the clusters to an initial relaxation process due to mixing during the formation of the clusters. However, the existence of 3 arbitrary parameters in formula (218) makes it possible to represent many density distributions with sufficient accuracy.

As yet there is no complete theory of the dissolution of the clusters that includes all the relevant factors. The main practical problem is to combine the effect discussed by Hénon (that encounters do not usually result in escapes of stars, but only cause some stars to move to the outer parts of the cluster), which tends to increase the lifetimes of the clusters, with the tidal effects that tend to reduce these lifetimes. Only when we have reliable lifetimes for the clusters we can compare them with observations, and derive cosmogonic consequences from them.

4. *The Evolution of Stellar Systems.* When one calculates the time of relaxation or the rate of escape of stars from a cluster, he usually assumes the density function as given. For example, Chandrasekhar assumes a constant density, King assumes a polytrope, etc. Therefore one assumes some "reasonable" model, found by means of the continuum approach of stellar dynamics, and then he proceeds to find the change of this model with time.

However, the problems of the structure and evolution of a cluster are not separate. The present structure of a cluster is the result of its evolution. Therefore the statistical approach of stellar dynamics should derive the distribution function f and its change in time from the interaction of stars among themselves. For example, one should prove a kind of Boltzmann's H-theorem (which states that any distribution of energy tends to a Boltzmann distribution, and the corresponding distribution of velocities tends to become Maxwellian).

These problems are not easy. We have not even succeeded in general in defining a probability density in the phase space of a stellar system. Further, a Boltzmann distribution corresponds to an infinite radius and mass. However there are some methods to escape from that difficulty.

First, we may consider the fact that the time scale of the evolution of a stellar system is different in different places. For example in the center of the observed clusters the time scale is comparatively short and a kind of statistical equilibrium has probably been reached. This equilibrium is slowly reaching the outer parts. However we may consider possible exceptions to this rule. Are all possible configurations of a cluster tending to the same equilibrium condition? It is quite probable that there are exceptional cases where such an equilibrium is never reached, even in the central parts of a cluster. Such are, for example, the cases where the motions of all the stars of the cluster are periodic. What happens then to systems that are "near" these exceptional cases? Do they reach an equilibrium, and in what time scale? For actual problems the calculation of a rough time scale of the evolution is quite necessary. If this time scale is extremely long we may consider a cluster as not changing at all.

Ogorodnikov [185], [212] considers stellar systems embedded in a greater stellar system that acts as a container. Such is the case of a Galaxy that is surrounded by a halo. The halo exchanges mass with the Galaxy and when a steady state is reached, the number of stars absorbed by the halo is equal to the number returned to the Galaxy. Then a probability density can be defined and a statistical mechanics of the Galaxy can be developed.

This approach is good for short time intervals. It is not completely satisfactory because we cannot consider the halo as a given boundary and as a matter of fact it is a part of the Galaxy itself. However, until a better theory of statistical mechanics of stellar systems is devised such approaches may be useful.

A quite different approach is due to Chandrasekhar and von Neumann [181], [213], [214], [215], [216], [217], [218]. They start from genuine statistical considerations; namely they introduce an a priori probability density for each star $\tau_i(\overline{r_i}, \overline{v_i})$, so that the probability that a star lies between $\overline{r_i}$ and $\overline{r_i} + d\overline{r_i}$ and has a velocity between $\overline{v_i}$ and $\overline{v_i} + d\overline{v_i}$ is $\tau_i dx_{i1}dx_{i2}dx_{i3}dv_{i1}dv_{i2}dv_{i3}$. Then the total probability density in the phase space is $T = \prod_{i=1}^{N} \tau_i$, where N is the number of stars.

Chandrasekhar and von Neumann used for τ_i the function

(219) $$\tau_i = \frac{j^3}{\pi^{3/2}} \exp(-j^2 v_i^2),$$

and more generally an arbitrary function of v_i^2 and of the mass m. However T is not an integral of the equations of motion (see e.g., R. Kurth [172]) because if

$$(220) \qquad T = \frac{j^3}{\pi^{3/2}} \exp(-j^2(v_1^2 + v_2^2 + \cdots + v_N^2)) = \text{constant},$$

then $v_1^2 + v_2^2 + \cdots + v_N^2 = $ constant, and this is not generally the case. Therefore this approach is not completely satisfactory, because the probability density changes along the orbit of each star.

However it has not been proved that a function τ_i with the required properties does not exist. On the other hand one may not be interested in individual stellar orbits at all. In fact in most cases we may consider a stellar system as a gas where the motions of the particles are subjected to perturbations quite similar to the Brownian motions of the particles of a fluid.

Chandrasekhar has calculated that the variations of the force acting on a star due to stellar encounters have a period of the order of 6×10^4 years in our Galaxy; therefore they are very fast in comparison with the period of galactic rotation of 2×10^8 years. Thus we may consider the force acting on a star as due to the smoothed out distribution of the stellar system plus a rapidly fluctuating force, as in the case of Brownian motion.

In order to describe the Brownian motion a generalization of Liouville's equation is usually used that is called the Fokker-Planck equation (see e.g., [218]). If f is the distribution function, its change in time because of the collisions is given by the equation

$$
\left(\frac{\partial f}{\partial t}\right)_c \Delta t + O(\Delta t^2) = -\sum_{i=1}^{3} \frac{\partial}{\partial \dot{x}_i}(f\langle \Delta \dot{x}_i \rangle)
$$

$$(221)$$

$$
+ \frac{1}{2} \sum_{i=1}^{3} \sum_{j=1}^{3} \frac{\partial^2}{\partial \dot{x}_i \partial \dot{x}_j}(f\langle \Delta \dot{x}_i \Delta \dot{x}_j \rangle)
$$

$$
+ O(\langle \Delta \dot{x}_i \Delta \dot{x}_j \Delta \dot{x}_k \rangle), \quad (k = 1, 2, 3),
$$

where $\langle \Delta \dot{x}_i \rangle$ is the mean value of $\Delta \dot{x}_i$ over the time Δt and is given by the formula

$$(222) \qquad \langle \Delta \dot{x}_i \rangle = \int_{-\infty}^{+\infty} \Delta \dot{x}_i \, \psi(\vec{x}; \Delta \vec{x}) \, d(\Delta \vec{x}),$$

where $\psi(\vec{\dot{x}}; \Delta\vec{\dot{x}})$ is the transition probability that $\vec{\dot{x}}$ will increase by $\Delta\vec{\dot{x}} \equiv (\Delta\dot{x}_1, \Delta\dot{x}_2, \Delta\dot{x}_3)$ in time Δt, and the integral is, in fact, triple. Similarly,

$$(223) \qquad \langle \Delta\dot{x}_i \Delta\dot{x}_j \rangle = \int_{-\infty}^{+\infty} \Delta\dot{x}_i \Delta\dot{x}_j \psi(\vec{\dot{x}}; \Delta\vec{\dot{x}}) \, d(\Delta\vec{\dot{x}}),$$

etc.

Liouville's equation is replaced by the generalized Fokker-Planck equation

$$(224) \qquad \frac{\partial f}{\partial t} + \sum_{i=1}^{3} \left[\frac{\partial f}{\partial x_i} \dot{x}_i - \frac{\partial f}{\partial \dot{x}_i} \frac{\partial V}{\partial x_i} \right] = \left(\frac{\partial f}{\partial t} \right)_c.$$

This equation corresponds to Boltzmann's equation in gases. If

$$(225) \quad \psi(\vec{\dot{x}}; \Delta\vec{\dot{x}}) = \frac{1}{(4\pi q \, \Delta t)^{3/2}} \exp\left\{ - | \Delta\vec{\dot{x}} + \beta\vec{\dot{x}} \Delta t |^2 / (4q \, \Delta t) \right\}.$$

(where the constant q is equal to $\beta kT/m$ in a gas, and $-\beta\vec{\dot{x}}$ is the dynamical friction), then $\langle \Delta\dot{x}_i \rangle = -\beta\dot{x}_i \Delta t$, $\langle \Delta\dot{x}_i \Delta\dot{x}_j \rangle = O(\Delta t^2)$ for $i \neq j$, and $\langle \Delta\dot{x}_i^2 \rangle = 2q\Delta t + O(\Delta t^2)$.

Hence

$$(226) \qquad \left(\frac{\partial f}{\partial t} \right)_c \Delta t + O(\Delta t^2) = \left\{ \beta \sum_{i=1}^{3} \frac{\partial (f\dot{x}_i)}{\partial \dot{x}_i} + q \sum_{i=1}^{3} \frac{\partial^2 f}{\partial \dot{x}_i^2} \right\} \Delta t;$$

i.e.,

$$(227) \qquad \left(\frac{\partial f}{\partial t} \right)_c = \beta \sum_{i=1}^{3} \frac{\partial (f\dot{x}_i)}{\partial \dot{x}_i} + q \sum_{i=1}^{3} \frac{\partial^2 f}{\partial \dot{x}_i^2}.$$

The Fokker-Planck equation now becomes

$$(228) \quad \frac{\partial f}{\partial t} + \sum_{i=1}^{3} \left[\frac{\partial f}{\partial x_i} \dot{x}_i - \frac{\partial f}{\partial \dot{x}_i} \frac{\partial V}{\partial x_i} \right] = \beta \sum_{i=1}^{3} \frac{\partial (f\dot{x}_i)}{\partial \dot{x}_i} + q \sum_{i=1}^{3} \frac{\partial^2 f}{\partial \dot{x}_i^2}.$$

Equations (224) and (228) have been used by a number of authors in discussing the dynamics of an ionized plasma (see [187], [219], [220], [221]). These problems are very similar to those of stellar dynamics.

A special application of Fokker-Planck's equation to stellar dynamics has been made by Chandrasekhar [195], [196], [197] in his study of dynamical friction. Another special case has been considered by Spitzer and Härm [189], in their calculation of the escape of stars from a cluster. A generalization of their solution was given by King [204].

A more general application of this equation was recently made by Hénon [222]. Hénon found a very important model of a stellar cluster that is homologous; i.e., this model is such that its evolution does not change the structure of the cluster but only its dimensions.

It is known that a stellar system cannot in general have a final equilibrium state, because the Maxwellian distribution implies an infinite radius and mass. And it is only in exceptional cases that a system may tend to a configuration where all the orbits are periodic. However, it may be that a system tends to a homologous configuration; i.e., its final evolution is homologous. Hénon has given examples where models initially different from the homologous model tend towards it. This is a case quite similar to the famous H-theorem of classical statistical mechanics, which states that the collisions tend to produce a Boltzmann distribution of the energy.

The homologous model found by Hénon has a finite mass and radius but its central density is infinite. However, the mass in any small volume around the origin is always finite. Hénon considered also some nonhomologous models which tend to the homologous model, and their central density tends to increase so that it becomes infinite in finite time.

In all these models the distribution function is a function of time and the energy integral only, $f = f(E, t)$. Further work should include the other integrals too.

A numerical solution of Fokker-Planck's equation for a special form of the distribution function, $f = f(E, C)$, including also the angular momentum integral, was given by Michie [223]. Michie also found a contraction of the core of the cluster, together with an expansion of its outer parts. Similar results were found lately by von Hörner [10] by means of numerical calculations of the n-body problem.

The effects of the encounters on the evolution of a stellar system may be divided into two classes; the effects due to close encounters (binary encounters), and those due to distant encounters. It is understood that the latter effects are more important than the former (see [187], [224]). A method that permits one to take into account not only binary encounters, but ternary and more complex encounters also, has been developed by Bogoliubov [225], Born and Green [226], [227], [228], [229], Kirkwood [230], [231], Yvon [232] and others [233], [234], especially in the case of a plasma. It should be very useful to develop this theory for stellar systems also. This problem is intimately connected with the problem of the time of relaxation. Further the change of the distribution function due to encounters $(\partial f/\partial t)_c$ in Fokker-Planck's equation may be derived more accurately by such a method.

As we have indicated above, the rate of ejection of stars from a cluster is not yet accurately known. The theory must explain quantitatively the extremely low escape rate indicated by von Hörner's numerical calculations [9], [10]. Further, the ejection of stars of different masses, the effects of the rotation and ellipticity of a cluster on the rate of ejection must be derived.

One very important problem is to find out if the stellar clusters tend always towards an homologous cluster.

Another problem is to find criteria that would indicate the age of a cluster; the present estimates of the ages of the clusters are in general based only on the evolution of different stellar models. It would be very important to have independent dynamical ages of the clusters.

There are many statistical problems concerning specific stellar systems, clusters, galaxies or clusters of galaxies. The statistical methods used in such problems are described, e.g., in the book of Trumpler and Weaver [110].

Statistical methods are introduced also in problems of celestial mechanics, in connection with the distribution of the asteroids, the comets and the artificial satellites around the earth (especially in the case of a great number of small satellites as e.g., the case of Project Westford). One recent example of this type, where a distribution function is used in celestial mechanics is given by Brouwer [235].

The most important problem of statistical mechanics of stellar systems at the present time is the justification of its basic assumptions and the development of new methods of attack on the basic statistical problems of stellar dynamics. This can be done with some success by numerical integrations of the n-body problem; the results reached this way will indicate the statistical methods to be developed in stellar dynamics. To this effect stellar dynamics has to cooperate with classical celestial mechanics on one hand plasma dynamics on the other. Much work is needed before we can construct a consistent and somewhat complete statistical theory of stellar systems.

I want to thank Dr. J. Barkley Rosser for suggesting many improvements in the text.

References

1. W. M. Smart, *Stellar dynamics,* Cambridge Univ. Press, Cambridge, 1938.

2. S. Chandrasekhar, *Principles of stellar dynamics,* Chicago Univ. Press, Chicago, Illinois, 1942; Dover, New York, 1960.

3. E. van der Pahlen, *Einführung in die Dynamik von Sternsystemen,* Birkhäuser Verlag, Basel, 1947.

4. R. Kurth, *Introduction to the dynamics of stellar systems,* Pergamon Press, New York, 1957.

5. B. Lindblad, *Galactic dynamics,* Handbuch der Physik **53** (1959), 21.

6. G. L. Camm, *Recent developments in stellar dynamics,* Vistas in Astronomy. I, (1955), 216.

7. K. F. Ogorodnikov, *Dynamics of stellar systems,* Moscow 1958; English transl. Pergamon Press, New York, 1965.

8. G. M. Idlis, *Structure and dynamics of stellar systems,* Alma Ata (1961).

9. S. von Hörner, Z. Astrophys. **50** (1960), 184.

10. _____ ,Z. Astrophys. **57** (1963), 47.

11. J. Chazy, Ann. Sci. École Norm. Sup., Sér. 3, **39** (1922), 29.

12. _____ , J. Math. Pures Appl., Sér. 9, 8 (1929), 353.

13. _____ , Bull. Astronom., Sér. 2, 8 (1932), 403.

14. G. F. Khilmi, *Qualitative methods in the many body problem,* Gordon and Breach, New York, 1961; J. F. Chilmi, *Qualitative methoden beim n-Körperproblem der Himmelsmechanik,* Akademie Verlag, Berlin, 1961.

15. *A review of Soviet celestial mechanics literature,* U.S. Department of Commerce, Office of Technical Services, Washington, 1958.

16. S. Chandrasekhar, J. Math. Anal. Appl. **1** (1960), 240.

17. _____ , Astrophys. J. **136** (1962), 1048.

18. S. Chandrasekhar and G. Contopoulos, Proc. Nat. Acad. Sci. U.S.A. **49**, (1963), 608.

19. M. Schwarzschild, Astronom. J. **59** (1954), 271.

20. V. A. Ambartsumian, *La structure et l'évolution de l'univers,* R. Stoops, Brussels, 1958; p. 256.

21. *Conference on the instability of systems of galaxies,* Astronom. J. **66** (1961), 533.

22. S. Chandrasekhar and N. R. Lebovitz, Astrophys. J. **135** (1962), 248.

23. _____ , Astrophys. J. **136** (1962), 1069,

24. U. V. Wijk, Ann. Astrophys. **12** (1949), 81.

25. I. King, Astronom. J. **66** (1961), 572.

26. N. Limber, Astrophys. J. **130** (1959), 414.

27. _____ , Astrophys. J. **135** (1962), 16.

28. _____ , Astrophys. J. **135** (1962), 41.

29. H. Poincaré, *Méthodes nouvelles de la mécanique céleste.* I, (1892); II, (1893); III, (1899), Gauthier-Villars, Paris; Dover, New York, 1961.

30. C. Carathéodory, Sitzungsber. Preuss. Akad. Wiss. 1919, p. 580.

31. E. Hopf, Math. Ann. 103 (1930), 710.

32. C. Siegel, *Vorlesungen über Himmelsmechanik,* Springer Verlag, Berlin, Göttingen, Heidelberg, 1956.

33. V. I. Arnol'd, Soviet Math. Dokl. 3 (1962), 1008.

34. K. Schwarzschild, Astronom. Nachr. 141 (1896), 1.

35. L. Becker, Monthly Notices Roy. Astronom. Soc. 80 (1920), 590.

36. _____, Monthly Notices Roy. Astonom. Soc. 80 (1920), 598.

37. J. Colleau, Séminaire Méc. Anal. Méc. Cél. (M. Janet) (9) 2 (1959).

38. R. Sibahara, Publ. Astronom. Soc. Japan 13 (1961), 108.

39. _____, Publ. Astronom. Soc. Japan 13 (1961), 113.

40. _____, Publ. Astronom. Soc. Japan 14 (1962), 10.

41. M. Hénon, Ann. Astrophys. 23 (1960), 668.

42. _____, Ann. Astrophys. 24 (1961), 369.

43. K. Sitnikov. Soviet Phys. Dokl. 5 (1961), 647.

44. S. von Hörner, Astrophys. J. 125 (1957), 451.

45. S. M. Ulam, Proc. 4th Berkeley Sympos. Math. Statistics and Probability, Vol. III, p. 315, Univ. of California Press, Berkeley and Los Angeles, Calif., 1961.

46. J. Boersma, B. A. Netherlands 15 (1961), 291.

47. A. Blaauw, B. A. Netherlands 15 (1961), 265.

48. J. H. Jeans, Monthly Notices Roy. Astronom. Soc., 76 (1915), 70.

49. H. Poincaré, *Leçons sur les hypothèses cosmogoniques,* Hermann, Paris, 1911.

50. S. Chandrasekhar, Astrophys. J. 90 (1939), 1.

51. _____, Astrophys. J. 92 (1940), 441.

52. M. Schürer, Astronom. Nachr. 273 (1943), 230.

53. G. L. Camm, Monthly Notices Roy. Astronom. Soc. 112 (1952), 155.

54. J. H. Jeans, *Problems of cosmogony and stellar dynamics,* Cambridge Univ. Press, Cambridge, 1919.

55. R. Kurth, Z. Astrophys. 26 (1949), 100.

56. _____, Z. Astrophys. 26 (1949), 168.

57. K. H. Prendergast, Astronom. J. 59 (1954), 260.

58. G. L. Camm, Monthly Notices Roy. Astronom. Soc. 110 (1950), 305.

59. R. Kurth, Astronom. Nachr. 282 (1955), 97.

60. R. Emden, *Gaskugeln,* Teubner Verlag, Leipzig, 1907.

61. D. N. Limber, Astrophys. J. 134 (1961), 537.

62. E. A. Kreiken, Ann. Astrophys. 24 (1961), 219.

63. _____, Ann. Astrophys. 25 (1962), 271.

64. G. L. Camm, Monthly Notices Roy. Astronom. Soc. 101 (1941), 195.

65. W. Fricke, Astronom. Nachr. 280 (1951), 125.

66. _____, Naturwiss. 38 (1951), 438.

67. D. Lynden-Bell, Monthly Notices Roy. Astronom. Soc. 123 (1962), 447.

68. E. Finlay-Freundlich and R. Kurth, Naturwiss. 7 (1955), 167.

69. J. H. Jeans, *Astronomy and cosmogony,* Cambridge Univ. Press, Cambridge, 1928; Dover, New York, 2nd 1961.

70. D. Lynden-Bell, Monthly Notices Roy. Astronom. Soc. 120 (1960), 204.

71. R. v.d. R. Woolley, Observatory 81 (1961), 161.

72. H. Nordström, Lund. Medd. II. No. 79 (1936).

73. A. Wintner, *Analytical foundations of celestial mechanics,* Princeton Univ. Press, Princeton, N.J., 1947.

74. G. Contopoulos, Astrophys. J. 138 (1963), 1297.

75. B. Lindblad, *Die Milchstrasse,* Handbuch der Astrophys. 5/2, (1933), 1047.
76. G. B. van Albada, Bosscha Contr. No. 1 (1952).
77. G. G. Kuzmin, Tartu Astr. Obs. Teated No. 1 (1953).
78. _____, Tartu Astr. Obs. Teated No. 2 (1956).
79. _____, Tartu Publ. 32 No. 5 (1953).
80. A. S. Eddington, Monthly Notices Roy. Astronom. Soc. 76 (1915), 37.
81. P. Stäckel, Math. Ann. 35 (1890), 91.
82. _____, Math. Ann. 42 (1893), 537.
83. J. Weinacht, Math. Ann. 90 (1923), 279.
84. H. C. van de Hulst, B. A. Netherlands 16 (1962), 235.
85. G. Contopoulos, Astronom. J. 63 (1963), 1.
86. D. Lynden-Bell, Monthly Notices Roy. Astronom. Soc. 124 (1962), 1.
87. _____, Monthly Notices Roy. Astronom. Soc. 124 (1962), 95.
88. E. T. Whittaker, Proc. Roy. Soc. Edinburgh 37 (1916), 95.
89. _____, *Analytical dynamics,* Cambridge Univ. Press, Cambridge, 1937; Dover, New York 1944.
90. T. M. Cherry, Proc. Cambridge Philos. Soc. 22 (1924), 325.
91. _____, Proc. Cambridge Philos. Soc. 22 (1924), 510.
92. _____, Proc. London Math. Soc., 2nd series 27 (1926), 151.
93. G. Contopoulos, Z. Astrophys. 49 (1960), 273.
94. G. D. Birkhoff, *Dynamical systems,* Amer. Math. Soc. Colloq. Publ. Vol. 9, Amer. Math. Soc., Providence, R. I., 1927.
95. D. Brouwer and G. M. Clemence, *Methods of celestial mechanics,* Academic Press, New York, 1961.
96. H. von Zeipel, Ark. Mat. Astr. Fys. 11 (1916), 1.
97. B. Barbanis, Z. Astrophys. 56 (1962), 56.
98. W. Gliese, Z. Astrophys. 39 (1956), 1; Mit. Astron. Rechen-Institut Serie A No. 8 (1957).
99. E. Strömgren, Astronom. Nachr. 203 (1917), 16.
100. G. Contopoulos, Z. Astrophys. 35 (1954), 67.
101. R. Kurth, Astronom. Nachr. 282 (1955), 241.
102. M. Hénon, Ann. Astrophys. 22 (1959), 491.
103. B. Lindblad, Monthly Notices Roy. Astronom. Soc. 94 (1934), 231; Stockholms Obs. Ann. (4) 11 (1935).
104. R. Coutrez, Ann. Obs. Roy. Belgique, 3 ème Sér. (Fasc. 3) 4 (1949).
105. J. Lense, Astronom. Nachr. 204 (1917), 17.
106. W. Gliese, Astronom. Nachr. 272 (1942), 201.
107. W. Lohmann, Z. Astrophys. 25 (1948), 293; ibid. 33 (1953), 186.
108. K. Schütte, Sitzungsber. Oester. Akad. Wiss. Math-Naturwiss. Kl. Abt. IIa, 161 (1952), 287.
109. M. Schwarzschild, Astronom. J. 57 (1952), 57.
110. R. J. Trumpler and H. F. Weaver, *Statistical astronomy,* Univ. of California Press, Berkeley and Los Angeles, Calif., 1953, p. 594ff.
111. I. Torgård, Medd. Lund Astr. Obs. Ser. II (133) 14 (1956).
112. U. van Wijk, Astronom. J. 61 (1956), 277.
113. M. C. Ballario, Oss. e. Mem. Arcetri. Fasc. 71 (1956).
114. S. Kikuchi, Sendaj Astr. Rap. No. 57 (1957).
115. _____, Sendaj Astr. Rap. No. 64 (1958).
116. R. M. Dzigvashvili, Bull. Abastumani Astroph. Obs. 19 (1955), 115; 23 (1958), 183.

117. S. Emoto, Publ. Astronom. Soc. Japan 10 (1958), 152.

118. L. Perek, Vistas in Astronomy 5 (1962), 28.

119. _____, B. A. Czeckoslovakia 13 (1962), 211.

120. B. Lindblad, Ark. Mat. Astr. Fys. (17) 20A (1927).

121. _____, Stockholms Obs. Ann. (4) 12 (1936).

122. _____, Festschrift für E. Strömgren, Einan Munksgaard, Kopenhagen, 1940, p. 131.

123. _____, Stockholms Obs. Ann. (10) 13 (1941).

124. B. Lindblad and F. Nahon, Stockholms Obs. Ann. (2) 18 (1954).

125. A. Blaauw, Publ. Groningen No. 52 (1946).

126. G. Contopoulos, Stockholms Obs. Ann. (10) 19 (1957).

127. T. Shimizu, Publ. Astronom. Soc. Japan 12 (1960), 238.

128. B. Lindblad, Stockholms Obs. Ann. (6) 18 (1955).

129. _____, Stockholms Obs. Ann. (7) 19 (1956).

130. _____, Stockholms Obs. Ann. (9) 19 (1957).

131. _____, Stockholms Obs. Ann. (6) 20 (1958).

132. _____, Stockholms Obs. Ann. (8) 21 (1961).

133. P. O. Lindblad, Stockholms. Obs. Ann. (4) 21 (1960).

134. M. Schmidt, B. A. Netherlands 13 (1956), 15.

135. G. Contopoulos, Stockholms Obs. Ann. (5) 20 (1958).

136. I. Torgård, Nuffic International Summer Course in Science, Part X, 1960.

137. A. Ollongren, B. A. Netherlands 16 (1962), 241.

138. G. Hori, Publ. Astronom. Soc. Japan 14 (1962), 353.

139. O. Eggen, Monthly Notices Roy. Astronom. Soc. 118 (1958), 65; ibid. 118 (1958), 154.

140. _____, Monthly Notices Roy. Astronom. Soc. 118 (1958), 560.

141. _____, Observatory 79 (1959), 143.

142. _____, Observatory 79 (1959), 197.

143. _____, Observatory 79 (1959), 206.

144. _____, Observatory 120 (1960), 430.

145. _____, Observatory 120 (1960), 448.

146. _____, Observatory 120 (1960), 540.

147. _____, Observatory 120 (1960), 563.

148. O. Eggen and A. P. Sandage, Monthly Notices Roy. Astronom. Soc. 119 (1959), 255.

149. G. Contopoulos and B. Barbanis, Observatory 82 (1962), 80.

150. R. v.d. R. Woolley, Vistas in Astronomy. III, (1960), 3.

151. _____, Observatory 81 (1961), 203.

152. I. King, Astrophys. J. 133 (1961), 347.

153. S. Chandrasekhar, Monthly Notices Roy. Astronom. Soc. 98 (1938), 710.

154. T. Shimizu, Publ. Astronom. Soc. Japan 14 (1962), 56.

155. S. Emoto, Publ. Astronom. Soc. Japan 14 (1962), 73.

156. G. Contopoulos, Astrophys. J. 124 (1956), 643.

157. A. Yasuda, Publ. Astronom. Soc. Japan 10 (1958), 165.

158. G. Contopoulos and G. Bozis, Contr. Astr. Dept. Univ. Thessaloniki No. 7 (1962).

159. S. P. Diliberto, W. T. Kyner and R. B. Freund, Astronom. J. 66 (1961), 118.

160. J. P. Vinti, J. Res. Nat. Bureau of Standards 63B (1959), 105; ibid. 65B (1961), 169.

161. G. Birkhoff, Proc. Nat. Acad. Sci. 17 (1931), 656.

162. A. Khinchin, *Mathematical foundations of statistical mechanics,* English transl., Dover, New York, 1949.

163. A. Münster, *Prinzipien der Statistischen Mechanik,* Handbuch der Physik. III, 2 (1959), 176.

164. K. Jacobs, *Neuere Methoden und Ergebnisse der Ergodentheorie,* Springer Verlag, Berlin, Göttingen, Heigelberg, 1960.

165. J. C. Oxtoby and S. M. Ulam, Ann. of Math. **42** (1941), 874.

166. S. P. Diliberto, *Perturbation theory of periodic surfaces.* III, Technical Report No. 10 (NR 041157), 1957.

167. J. Moser, Nachr. Akad. Wiss. Göttingen, Math. Phys. Kl. 1 (1962).

168. _____, *Nonlinear problems,* ed. R. E. Langer, Univ. of Wisconsin Press, Madison, Wisc., 1963, p. 139.

169. V. I. Arnol'd, Soviet Math. Dokl. **2** (1961), 247.

170. _____, Soviet Math. Dokl. **3** (1962), 136.

171. C. Goudas and B. Barbanis, Z. Astrophys. **57** (1963), 183.

172. R. Kurth, Zeitschrift f. Angew. Math. Phys. **6** (1955), 115.

173. O. Heckmann, Z. Astrophys. **23** (1944), 31.

174. P. and T. Ehrenfest, *The conceptual foundation of the statistical approach in mechanics,* English transl., Cornell Univ. Press, Ithaca, N. Y., 1959.

175. S. Böhme, Naturwiss. **43** (1956), 189.

176. W. Fricke, Z. Astrophys. **19** (1940), 304.

177. J. Belzer, L. Gamow and G. Keller, Astrophys. J. **113** (1951), 166.

178. S. von Hörner, Astrophys. J. **125** (1957), 451.

179. W. Fricke, Z. Astrophys. **20** (1941), 268.

180. O. Heckmann, *Astronomie, Astrophysik and Kosmogonie,* ed. P. ten Bruggen-cate, Dieterichsche Verlagsbuchhandlung, Wiesbaden, 1948.

181. S. Chandrasekhar, Astrophys. J. **94** (1941), 511.

182. L. Spitzer and M. Schwarzschild, Astrophys. J. **114** (1951), 385.

183. _____, Astrophys. J. **118** (1953), 106.

184. D. Osterbrock, Astrophys. J. **116** (1952), 164.

185. K. V. Ogorodnikov, Soviet Astr. 1 (1959), 787.

186. M. Hénon, Ann. Astrophys. **21** (1958), 186.

187. R. S. Cohen, L. Spitzer and P. M. Routly, Phys. Rev. **80** (1950), 230.

188. D. Pines and D. Bohm, Phys. Rev. **85** (1952), 338.

189. L. Spitzer and R. Härm, Astrophys. J. **127** (1958), 544.

190. D. Lynden-Bell, Monthly Notices Roy. Astronom. Soc. **124** (1962), 23.

191. D. Layzer, Astrophys. J. **137** (1963), 351.

192. I. Langmuir, Phys. Rev. **26** (1925), 585.

193. _____, Z. Phys. **46** (1928), 271.

194. D. Gabor, E. A. Ash and D. Dracott, Nature **176** (1955), 116.

195. S. Chandrasekhar, Astrophys. J. **97** (1943), 255.

196. _____, Astrophys. J. **97** (1943), 263.

197. _____, Astrophys. J. **98** (1943), 54.

198. M. L. White, Astrophys. J. **109** (1949), 159.

199. I. King, Astronom. J. **64** (1959), 351.

200. R. v.d. R. Woolley and A. Robertson, Monthly Notices Roy. Astronom. Soc. **116** (1956), 288.

201. I. King, Astronom. J. **63** (1958), 109.

202. _____, Astronom. J. **63** (1958), 114.

203. _____, Astronom. J. **63** (1958), 465.

204. ———, Astronom. J. **65** (1960), 122.

205. S. von Hörner, Z. Astrophys. **44** (1958), 221.

206. L. Spitzer, Monthly Notices Roy. Astronom. Soc. **100** (1940), 396.

207. R. v.d. R. Woolley and R. J. Dickens, Roy. Obs. Bull. No. 54 (1962).

208. B. Bok, Harvard Circ. No. 384 (1934).

209. H. Mineur, Ann. Astrophys. **2** (1939), 1.

210. I. King, Astronom. J. **67** (1962), 471.

211. L. Spitzer, Astrophys. J. **127** (1958), 17.

212. K. V. Ogorodnikov, Soviet Astr. **1** (1959), 748.

213. S. Chandrasekhar and J. von Neumann, Astrophys. J. **95** (1942), 489.

214. ———, Astrophys. J. **97** (1943), 1.

215. ———, Astrophys. J. **97** (1943), 25.

216. ———, Astrophys. J. **99** (1944), 47.

217. S. Chandrasekhar, Ann. New York Acad. Sci. **45** (1943), 131.

218. ———, Rev. Mod. Phys. **15** (1943), 1.

219. L. Spitzer and R. Härm, Phys. Rev. **89** (1953), 977.

220. S. Gasiorowicz, M. Neuman and R. J. Riddell, Jr., Phys. Rev. **101** (1956), 922.

221. M. N. Rosenbluth, W. M. MacDonald and D. L. Judd, Phys. Rev. **107** (1957), 1.

222. M. Hénon, Ann. Astrophys. **24** (1961), 369.

223. R. W. Michie, Astrophys. J. **133** (1961), 781.

224. S. Chapman and T. G. Cowling, *The mathematical theory of non-uniform gases,* Cambridge Univ. Press, Cambridge 1960. p. 178.

225. N. Bogoliubov, *Problems of a dynamical theory in statistical physics,* English transl., Geophysics Research Directorate, USAF, Bedford, Mass., 1960.

226. M. Born and H. S. Green, Proc. Roy. Soc. **A188** (1946), 10.

227. H. S. Green, Proc. Roy. Soc. **A189** (1947), 103.

228. M. Born and H. S. Green, Proc. Roy. Soc. **A190** (1947), 455.

229. ———, *A general kinetic theory of liquids,* Cambridge Univ. Press, Cambridge, 1949.

230. J. G. Kirkwood, J. Chem. Phys. **14** (1946), 180.

231. ———, J. Chem. Phys. **15** (1946), 72.

232. J. Yvon, *La theorie statistique der fluides et l'équation d'état,* Hermann, Paris, 1935.

233. C. M. Tchen, Phys. Rev. **114** (1959), 394.

234. N. Rostoker and M. N. Rosenbluth, *The physics of fluids,* **3** (1960), 1.

235. D. Brouwer, Astronom. J. **68** (1963), 152.

236. P. Bouvier, Publ. Obs. Genève A **63** (1963), 163.

237. G. Contopoulos, Astronom. J. **68** (1963), 763.

UNIVERSITY OF THESSALONIKI
THESSALONIKI, GREECE

Harry Pollard

Qualitative Methods
in the n-Body Problem[*]

I. **Introduction.** The n-body problem is concerned with the motion of masses $m_1, \cdots, m_n$ $(n > 1)$, moving in inertial space under the attraction of their gravitational forces. In the case of a particle m_k being acted upon by a mass m_j, we illustrate the geometry in Figure 1.

With position vectors $\mathbf{r}_j$ and $\mathbf{r}_k$, the differential equation of motion due to the force on the kth particle by the other masses is

$$(\text{I.1}) \qquad m_k \ddot{\mathbf{r}}_k = \sum_{j=1; j \neq k}^{n} \gamma \frac{m_j m_k}{r_{jk}^2} \frac{\mathbf{r}_j - \mathbf{r}_k}{r_{jk}} \qquad (k = 1, \cdots, n).$$

Assuming that the initial position and velocity are given, i.e., $\mathbf{r}_k(0)$, $\mathbf{v}_k(0)$ and $r_{jk} > 0$, we seek a solution of (I.1).

To realize what constitutes a solution to a differential equation, recall the problem

$$\frac{dy}{dx} = f(x, y),$$

where we seek a solution passing through a predetermined point

---

[*] This is a reporter's account of Professor Pollard's lectures, partially revised by the editor.

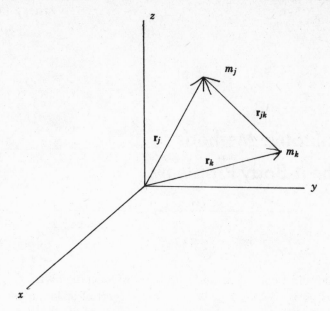

FIGURE 1. Two Interacting Masses

(x_0, y_0). In general $y = g(x, c)$ is said to be a solution depending on the parameter c if

$$\frac{\partial}{\partial x} g(x, c) \equiv f(x, g(x, c)).$$

Obviously to find c we solve $y_0 = g(x_0, c)$.

More generally, we may say that $h(x, y, c) = 0$ is the solution, meaning that if we determine $g(x, c)$ implicitly by $h(x, g(x, c), c) \equiv 0$, then $y = g(x, c)$ is the solution in the earlier sense. For example consider

$$\frac{dy}{dx} = \frac{2x + ye^{xy}\cos e^{xy}}{xe^{xy}\sin^{xy} + 1},$$

with solution

$$x^2 + \sin e^{xy} + y = c.$$

The latter equation is a solution in the sense that if it is differentiated the former equation results. Actually such a solution serves

little useful purpose unless there exists some transparency that
makes it more tractable than the example just given.

Consider a set of differential equations

$$\frac{dx}{dt} = f(x,y), \quad \frac{dy}{dt} = g(x,y)$$

with initial conditions $x(0)$ and $y(0)$ given. The problem is to find
solutions $x = x(t)$ and $y = y(t)$ satisfying the differential equations
and the initial conditions. Simple division of these equations eliminates the variable t and yields $dy/dx = h(x,y)$ where $y = g(x)$. Here
we have managed to reduce the system by one, and there is a chance
that if the solution is transparent the reduction is useful. In general
it is not. Thus mathematicians were led to look for integrals to
systems of differential equations. Returning to equation (I.1), the
idea is to reduce it to a system of first-order differential equations
of the form

(I.2) $$\frac{dx_k}{dt} = f_k(x_1, \cdots, x_m) \qquad (k = 1, \cdots, m),$$

where $x_k = x_k(t)$ and $x_k(0)$ is given for $k = 1, \cdots, m$. The order of (I.2)
is m, with $m = 6n$. A function $g(x_1, \cdots, x_m, t)$ is called an integral of
the system if every solution of the system gives

$$g(x_1(t), \cdots, x_m(t), t) = \text{constant},$$

where the constant is determined by

$$g(x_1(t), \cdots, x_m(t), t) = g(\text{initial values}).$$

If we can find m integrals

(I.3) $g_k(x_1(t), \cdots, x_m(t), t) = g_k(\text{initial values}) \qquad (k = 1, \cdots, m)$

they constitute an implicit solution of (I.2) in that they are m equations in m unknowns for which we can solve $x_k = x_k(t, \text{initial condi-}$
tions), and the problem is solved in terms of t and the initial conditions.

II. **Illustrative central force problem.** To treat the two-body problem, $n = 2$, $m = 12$, with masses moving in a field subject to the inverse square law, we first look at a central force problem for one
body, $n = 1$, $m = 6$,

$$m\ddot{\mathbf{r}} = -\frac{\mu m}{r^2}\frac{\mathbf{r}}{r},$$

(II.1)   or

$$\ddot{\mathbf{r}} = -\frac{\mu}{r^3}\mathbf{r},$$

where $r = |\mathbf{r}|$.

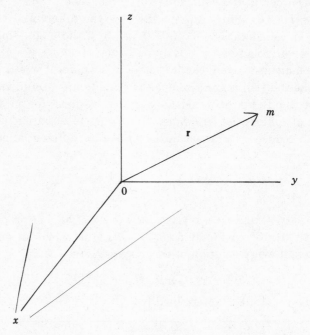

FIGURE 2. A Single Body with a Central Force

Using Laplace's method, we write

(II.2) $$\frac{d}{dt}\frac{\mathbf{r}}{r} = \frac{r\mathbf{v} - r\dot{r}}{r_1^2}.$$

Recalling that $r^2 = \mathbf{r}^2$ and $r\dot{r} = \mathbf{r}\cdot\dot{\mathbf{r}}$, (II.2) can be written as

(II.3) $$\frac{d}{dt}\frac{\mathbf{r}}{r} = \frac{(\mathbf{r}\times\mathbf{v})\times\mathbf{r}}{r^3} = -\frac{(\mathbf{r}\times\mathbf{v})\times\ddot{\mathbf{r}}}{\mu}.$$

Since $\mathbf{r}\times\ddot{\mathbf{r}} = 0$ by (II.1), we have $\mathbf{r}\times\mathbf{v} = h$, which constitutes three integrals. Using this in (II.3) gives

(II.4) $$\frac{d}{dt}\frac{\mathbf{r}}{r} = \frac{-\mathbf{h}\times\ddot{\mathbf{r}}}{\mu}.$$

Integrating (II.4), we have

(II.5)
$$\frac{\mathbf{r}}{r} = \frac{-\mathbf{h} \times \mathbf{v}}{\mu} - \mathbf{e},$$

$$\frac{\mathbf{r}}{r} + \mathbf{e} = \frac{\mathbf{v} \times \mathbf{h}}{\mu}$$

which constitutes three more integrals. However, the problem is not complete since (I.2) has at least one solution of the form

$$g_k(x_1, \cdots, x_m, t) = \text{constant},$$

in which t appears explicitly. But the 6 integrals shown above contain no such function, implying that $\mathbf{e}$ and $\mathbf{h}$ are not independent of each other. In fact

$$\mathbf{e} \cdot \mathbf{h} = 0.$$

Thus, in fact, (II.5) yields but 2 new integrals, with the sixth, called the time of perihelion passage, still missing.

Turning to the larger problem of two bodies, initially at different points, we seek 12 integrals for the system (see (I.1) and Figure 1):

(II.6)
$$m_1 \ddot{\mathbf{r}}_1 = \frac{-\gamma\, m_1 m_2}{r_{12}^2} \frac{(\mathbf{r}_1 - \mathbf{r}_2)}{r_{12}},$$

(II.7)
$$m_2 \ddot{\mathbf{r}}_2 = \frac{-\gamma\, m_1 m_2}{r_{12}^2} \frac{(\mathbf{r}_2 - \mathbf{r}_1)}{r_{12}}.$$

Adding (II.6) and (II.7) yields

(II.8)
$$m_1 \ddot{\mathbf{r}}_1 + m_2 \ddot{\mathbf{r}}_2 = 0.$$

Define $M = m_1 + m_2$ and $\mathbf{r}_c$ (center of mass) $= M^{-1}(m_1 \mathbf{r}_1 + m_2 \mathbf{r}_2)$, so that $\ddot{\mathbf{r}}_c = 0$. This last equation indicates the center of mass is not accelerating. Integrating to get the velocity and position of the center of mass,

$\mathbf{v}_c = \mathbf{l}$ (3 integrals, conservation of linear momentum),

$\mathbf{r}_c = \mathbf{l}t + \mathbf{j}$ (3 integrals).

Multiplying (II.6) by $\times \mathbf{r}_1$ and (II.7) by $\times \mathbf{r}_2$ and adding, we get

$$m_1(\mathbf{r}_1 \times \ddot{\mathbf{r}}_1) + m_2(\mathbf{r}_2 \times \ddot{\mathbf{r}}_2) = 0.$$

Integration yields

(II.9) $\qquad m_1(\mathbf{r}_1 \times \mathbf{v}_1) + m_2(\mathbf{r}_2 \times \mathbf{v}_2) = \mathbf{h}.$

These are three more integrals, expressing the constancy of angular momentum. Now multiply (II.6) by $\cdot \dot{\mathbf{r}}_1$, (II.7) by $\cdot \dot{\mathbf{r}}_2$ and add.

$$m_1 \dot{\mathbf{r}}_1 \cdot \ddot{\mathbf{r}}_1 + m_2 \dot{\mathbf{r}}_2 \cdot \ddot{\mathbf{r}}_2 = \frac{d}{dt} \frac{\gamma m_1 m_2}{r_{12}}.$$

Integrating, we have

(II.10) $\qquad \dfrac{1}{2}(m_1 v_1^2 + m_2 v_2^2) = \dfrac{\gamma m_1 m_2}{r_{12}} + E.$

This is an integral expressing the conservation of energy.

Subtracting (II.6) from (II.7) to get the equation of motion of the second particle with respect to the first, we have

(II.11) $\qquad \ddot{\mathbf{r}}_2 - \ddot{\mathbf{r}}_1 = -\dfrac{\gamma(m_1 + m_2)}{r_{12}^3}(\mathbf{r}_2 - \mathbf{r}_1).$

Let $\mathbf{r} = \mathbf{r}_2 - \mathbf{r}_1$ and $\mu = \gamma M$ so that (II.11) becomes

(II.12) $\qquad \ddot{\mathbf{r}} = -\dfrac{\mu}{r^3}\,\mathbf{r}.$

This is the central force problem considered above, which yields five more integrals. Since we now have more than 12 integrals, some must be redundant. They can be reduced to the following:

$$\mathbf{v}_c = \mathbf{l}, \qquad\qquad (3)$$

$$\mathbf{r}_c = \mathbf{l}t + \mathbf{j}, \qquad\qquad (3)$$

(II.13) $\qquad m_1(\mathbf{r}_1 \times \mathbf{v}_1) + m_2(\mathbf{r}_2 \times \mathbf{v}_2) = \mathbf{h}, \qquad (3)$

$$\frac{1}{2}(m_1 v_1^2 + m_2 v_2^2) = \frac{v m_1^4 m_2}{r_{12}} + E, \qquad (1)$$

$$\frac{\mathbf{r}}{r} + \mathbf{e} = \frac{\mathbf{v} \times \mathbf{h}}{\mu}. \qquad\qquad (2)$$

Returning to the problem of the time of perihelion passage, operating on both sides of the last equation in (II.13) by $\cdot \mathbf{r}$ we have

$$r + \mathbf{e} \cdot \mathbf{r} = \frac{h^2}{\mu},$$

which can be rewritten as

(II.14)
$$r = \frac{h^2}{\mu(1 + e \cos \omega)}$$

which is the polar equation of a conic section with major axis along e, and ω the angle between e and r.

Finally, squaring the last equation in (II.13) we have

$$1 + \frac{2}{r} \left\{ \frac{h^2}{\mu} - r \right\} + e^2 = \frac{v^2 h^2}{\mu^2},$$

$$\frac{v^2}{2} = \frac{\mu}{r} + \frac{(e^2 - 1)\mu^2}{2h^2} = \frac{\mu}{r} + E.$$

This is the conservation of energy statement, with hyperbolic motion for $e^2 > 1$, parabolic for $e^2 = 1$, elliptic for $e^2 < 1$; for hyperbolic or elliptic motion, we must have $h \neq 0$.

Since $|\mathbf{a} \cdot \mathbf{b}|^2 + |\mathbf{a} \times \mathbf{b}|^2 = a^2 b^2$, substituting r and v we have $r^2 \dot{r}^2 + h^2 = r^2 v^2$ or

(II.15)
$$\frac{1}{2} \left(\dot{r}^2 + \frac{h^2}{r^2} \right) = \frac{\mu}{r} + E.$$

But $v^2/2 = (\mu/r) + E$ is valid even if $h = 0$, since if $h = 0$, then

$$\ddot{r} = -\mu/r^2.$$

Without integrating, (II.15) shows

(II.16)
$$\frac{1}{2} \frac{h^2}{r^2} \leq \frac{\mu}{r} + E.$$

Multiplying (II.16) by r^2, we have

(II.17)
$$\frac{1}{2} h^2 \leq \mu r + E r^2.$$

From (II.17) we see that if $r \to 0$, then $h = 0$. Analogously, in the n-body problem, all bodies cannot collide simultaneously unless the total angular momentum is 0.

Multiplying (II.15) by $2r^2$ and simplifying, we have

(II.18)
$$(r\dot{r})^2 + h^2 = 2(\mu r + E r^2).$$

In intervals when $r \neq 0$, we can introduce a variable u such that $\dot{u} = k/r$. Then

$$\dot{r} = \frac{dr}{d\mu}\frac{d\mu}{dt}, \, r\dot{r} = kr',$$

if we let a prime denote differentiation with respect to u. Then (II.18) becomes

(II.19) $$k^2(r')^2 + h^2 = 2(\mu r + Er^2).$$

Differentiating (II.19) and dividing by $2r'$ yields

(II.20) $$k^2 r'' = \mu + 2Er.$$

For the case where $E = 0$ in (II.20), choose $k^2 = \mu$, so that $r'' = 1$ and

(II.21) $$r = \frac{u^2}{2} + au + b.$$

As $t = 1/k \int r \, du$, we get

(II.22) $$t = u^{-1/2}\left(\frac{u^3}{6} + \frac{au^2}{2} + bu + c\right)$$

and we obviously have parabolic motion.

For $E < 0$, choose $k^2 = 2|E|$, yielding $r'' + r = \mu/k^2$. Thus $r = \mu/k^2 + A\cos(u - B)$.

By proper choice of the constant of integration in

$$u = \int \frac{k}{r} \, dt$$

we can take $B = 0$. So

(II.23) $$r = \frac{\mu}{k^2} + A\cos u,$$

(II.24) $$t = t_0 + \frac{\mu}{k^3} u + \frac{A}{k}\sin u.$$

For $E > 0$, choose $k^2 = 2E$, yielding

(II.25) $$r = \frac{\mu}{k^2} + A\cosh u,$$

(II.26) $$t = t_0 + \frac{\mu}{k^3} u + \frac{A}{k}\sinh u.$$

Let us now discuss two basic problems of interest in the two-body problem:

(1) For those orbits in which the masses are separating as $t \to \infty$, how large is r?

(2) If $h = 0$, and for some time $t = t_1$, a collision occurs, how small is r?

For problem (1) we obtain from

(II.21) and (II.22), $r \sim t^{2/3}$, for $E = 0$,

(II.25) and (II.26), $r \sim t$, for $E > 0$,

(II.23) and (II.24), r bounded, for $E < 0$.

Now consider the problem when $h = 0$ and collision occurs at $t = t_1$. In short, in what way is r related to $(t_1 - t)$ as $t \to t_1$. From $\ddot{x} = -\mu/x^2$, multiplying by $\dot{x}$ and integrating, we have $\dot{x}^2/2 = \mu/x + E$. Multiplying now by x and taking the limit as $x \to 0$ or $t \to t_1$, we find

(II.27) $$\lim x\dot{x}^2 = 2\mu.$$

Assume there exists an α such that $x \sim (t_1 - t)^\alpha$. Substitution in (II.27) yields

$$\alpha^2 (t - t_1)^\alpha (t_1 - t)^{2\alpha - 2} \to 2\mu,$$

$$(t_1 - t)^{3\alpha - 2} \to 2\mu/\alpha^2,$$

$$x \sim (t_1 - t)^{2/3} \text{ as } t \to t_1.$$

It is easy to make this result rigorous.

III. **Introduction to the n-body problem.** An interesting property (see Bertrand, [1]) is that if one has a particle in a circular orbit and the initial conditions are changed slightly, only the inverse square law (μ/r^2) and the linear law (μr) will yield a new closed orbit. In the solar system under a linear law, the planets would move in elliptic orbits with the sun at the center, and with a common period.

Let us first discuss the n-body problem under an arbitrary law $f(r)$. The equations of motion become

(III.1) $$m_k \ddot{\mathbf{r}}_k = \sum_{j=1; j \neq k}^{n} m_j m_k f(r_{jk}) \frac{\mathbf{r}_j - \mathbf{r}_k}{r_{jk}}.$$

Summing (III.1) over all values of k,

(III.2) $$\sum_{k=1}^{n} m_k \ddot{\mathbf{r}}_k = 0.$$

Again using the concept of the mass center, with $M = m_1 + m_2 + \cdots + m_n$,

$$\mathbf{r}_c = \frac{1}{M} (m_1 \mathbf{r}_1 + m_2 \mathbf{r}_2 + \cdots),$$

we get

$$\ddot{\mathbf{r}}_c = 0,$$

$$\dot{\mathbf{r}}_c = \mathbf{l},$$

$$\mathbf{r}_c = \mathbf{l}t + \mathbf{j}.$$

It simplifies matters if we change to a new coordinate system in constant linear motion with respect to the first in which

(III.3) $$\ddot{\mathbf{r}}_c = \dot{\mathbf{r}}_c = \mathbf{r}_c = 0.$$

The equations (III.1) are unchanged, so we have

(III.4) $$\sum_{k=1}^{n} m_k \mathbf{r}_k = 0,$$

(III.5) $$\sum_{k=1}^{n} m_k \mathbf{v}_k = 0.$$

The order of the system has now been reduced to $6n - 6$.

If $f(r) = \gamma r$, (III.1) can be reduced to

(III.6) $$\ddot{\mathbf{r}}_k = - \gamma M \mathbf{r}_k, \qquad k = 1, \cdots, n.$$

(III.6) implies all the masses satisfy the same differential equation, but fails to recognize that perhaps two of the masses may collide. From (III.6), the motion is elliptic or linear, and

(III.7) $$\mathbf{r}_k = \mathbf{A}_k \cos \omega t + \mathbf{B}_k \sin \omega t, \omega = \sqrt{(\gamma M)}.$$

For the situation $f(r) = \gamma/r^2$ where $n = 2$ or 3, if the solution to (III.1) ceases to be analytic at some time $t = t_1$, a collision has occurred. For $n > 3$ the problem remains unsolved.

However, Painlevé has shown (see [3]) that if in some finite time $t = t_1$, a singularity occurs, then

$$\min_{t \to t_1} r_{jk} = 0,$$

where we have $n(n-1)/2$ distances r_{jk}.

Returning to (III.1), crossing by $\mathbf{r}_k$ and integrating we get

(III.8) $$\sum_{k=1}^{n} m_k(\mathbf{r}_k \times \mathbf{v}_k) = \mathbf{h},$$

subject to (III.4) and (III.5). The system has now been reduced to $6n - 9$.

For the final reduction, define $\mu(r)$ such that $\mu'(r) = -f(r)$. Form the self potential

(III.9) $$U = \sum_{1 \le j < k \le n} m_j m_k \mu(r_{jk}).$$

Relating (III.9) to (III.1), we have

(III.10) $$m_k \ddot{\mathbf{r}}_k = \text{grad}_k U.$$

Multiplying (III.10) by $\dot{\mathbf{r}}_k$ and integrating,

(III.11) $$\frac{1}{2} \sum_{k=1}^{n} m_k v_k^2 = E + U.$$

This embodies the conservation of energy.

To specify the constant, and reduce the system to $6n - 10$, for

$$f(r) = \frac{\gamma}{r^2}, \text{ let } \mu(r) = \int_r^{\infty} f(r)\, dr,$$

$$f(r) = \gamma r, \text{ let } \mu(r) = -\int_0^r f(r)\, dr,$$

$$f(r) = \frac{\gamma}{r}, \text{ let } \mu(r) = -\int_1^r f(r)\, dr.$$

IV. **The Lagrange-Jacobi identity.** Let us define I by

(IV.1) $$I = \frac{1}{2} \sum_{k=1}^{n} m_k r_k^2.$$

Observe that this is half the usual moment of inertia. Since

$$2I = \sum_{k=1}^{n} m_k \mathbf{r}_k \cdot \mathbf{r}_k$$

it follows that

(IV.2)
$$\dot{I} = \sum_{k=1}^{n} m_k \mathbf{r}_k \cdot \mathbf{v}_k.$$

Therefore

(IV.3)
$$\ddot{I} = \sum_{k=1}^{n} m_k (\mathbf{r}_k \cdot \ddot{\mathbf{r}}_k + \mathbf{v}_k^2)$$

$$= 2T + \sum_{k=1}^{n} m_k \mathbf{r}_k \cdot \ddot{\mathbf{r}}_k.$$

But $\sum_{k=1}^{n} m_k \mathbf{r}_k \cdot \ddot{\mathbf{r}}_k = \sum_{k=1}^{n} \mathbf{r}_k \cdot \mathrm{grad}_k U$, so we have

(IV.4)
$$\ddot{I} = 2T + \sum_{k=1}^{n} \mathbf{r}_k \cdot \mathrm{grad}_k U.$$

A function $f(x_1, \cdots, x_m)$ is homogeneous of order k if for $0 < \lambda$

(IV.5)
$$f(\lambda x_1, \cdots, \lambda x_m) = \lambda^k f(x_1, \cdots, x_m).$$

Differentiating (IV.5) with respect to λ, and letting λ become 1 we get

(IV.6)
$$\sum_{s=1}^{n} x_s \frac{\partial f}{\partial x_s} = kf.$$

This is precisely the expression for the coordinates in (IV.4) if U is homogeneous of order l. i.e.,

(IV.7)
$$\sum_{k=1}^{n} \left(X_k \frac{\partial U}{\partial X_k} + Y_k \frac{\partial U}{\partial Y_k} + Z_k \frac{\partial U}{\partial Z_k} \right) = lU.$$

Thus

(IV.8)
$$\ddot{I} = 2T + lU.$$

This is known as the Lagrange-Jacobi identity. Consider the effect of letting $f(r) = \gamma/r^p$, $-\infty < p < \infty$.

In the previous section, $\mu(r)$ was defined as follows:

$$\mu(r) = \int f(r)dr,$$

with the following choice of limits.

$$\text{If } p > 1, \quad \mu(r) = \gamma \int_r^\infty \frac{dr}{r^p} = \frac{1}{p-1} \cdot \gamma r^{1-p}.$$

$$\text{If } p < 1, \quad \mu(r) = -\gamma \int_0^r \frac{dr}{r^p} = \frac{1}{p-1} \cdot \gamma r^{1-p}.$$

$$\text{If } p = 1, \quad \mu(r) = -\gamma \int_1^r \frac{dr}{r} = \gamma \log(1/r).$$

Thus, if $p < 1$ or $p > 1$, $\mu(r)$ and consequently $U(r)$ is homogeneous of degree $(1-p)$. (Recall (III.9).)

Applying the homogeneity property to (IV.8),

(IV.9) $$\ddot{I} = 2T + (1-p)U$$

for $f(r) = \gamma r^{-p}$, $p \neq 1$.

Since by (III.11) we have $T = U + E$ and $U = T - E$, (IV.9) becomes

(IV.10)
$$\ddot{I} = 2U + 2E + (1-p)U$$
$$= (3-p)U + 2E \equiv (3-p)T + (p-1)E.$$

We are now in a position to discuss qualitatively the relationship between I, U, T and the general geometry of the problem.

Let O be the mass center of the three-body system shown in Figure 3, and define

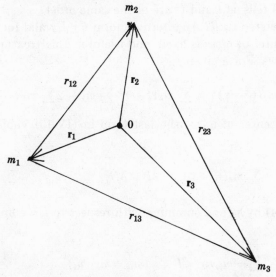

FIGURE 3. Three Interacting Masses

$$R(t) = \max r_{jk}(t),$$

$$\zeta(t) = \max r_k(t),$$

$$r(t) = \min r_{jk}(t),$$

where these functions by their very definition are not necessarily analytic. The use of inequalities will enable us to relate I to these new functions. By (IV.1)

$$2I = \sum_{k=1}^{n} m_k r_k^2.$$

By definition, each $r_k^2 \leq \zeta^2(t)$

$$\therefore 2I \leq \sum_{k=1}^{n} m_k \zeta^2,$$

or

(IV.11) $$I \leq \frac{M}{2} \zeta^2(t).$$

Similarly, $\sum m_k r_k^2 \geq m \sum r_k^2 \geq m\zeta^2$ where $m = \min m_k$. Combining this result with (IV.11),

(IV.12) $$\frac{m}{2} \zeta^2 \leq I \leq \frac{M}{2} \zeta^2.$$

Inequality (IV.12) tells us I and ζ^2 are of the same order.

Now how is I related to R? An alternate form for I, valid for the case where the center of mass is fixed, and useful in this treatment, is derived as follows. For a given j

(IV.13) $$\sum_{k=1}^{n} m_k(\mathbf{r}_k - \mathbf{r}_j)^2 = \sum_{k=1}^{n} m_k r_k^2 + r_j^2 \sum_{k=1}^{n} m_k - 2\sum_{k=1}^{n} m_k \mathbf{r}_k \cdot \mathbf{r}_j.$$

But with a fixed center of mass, the last term in (IV.13) vanishes by (III.4), and

(IV.14) $$\sum_{k=1}^{n} m_k(\mathbf{r}_k - \mathbf{r}_j)^2 = 2I + Mr_j^2.$$

Multiplying (IV.14) by m_j and summing with respect to j, we finally get

(IV.15) $$\frac{1}{2M} \sum_{1 \leq j < k \leq n} m_{jk} r_{jk}^2 = I \qquad (m_{jk} = m_j \cdot m_k).$$

But from the definition of $R(t)$,

$$I \leq \frac{1}{2M} R^2(t) \sum m_{jk} \text{ or simply}$$

(IV.16)

$$I \leq AR^2(t) \text{ where } A \text{ is a constant.}$$

Treating (IV.15) as we treated (IV.1) to get (IV.12), we get constants A and B such that

(IV.17) $$BR^2(t) \leq I \leq AR^2(t),$$

so I is of the order of $R^2(t)$. A general conclusion is that if for $t \to a$ one of the quantities I, R or $\varsigma \to \infty$, they all $\to \infty$.

Let us now show the relation between r, U and T for $1 < p < 3$. Since $r_{jk} \geq r$, $1/r_{jk} \leq 1/r$, and from (III.9),

(IV.18) $$\frac{B}{r^{p-1}} \leq U = \sum_{i \leq j < k \leq n} \frac{\gamma m_{jk}}{(p-1)r_{jk}^{p-1}} \leq \frac{A}{r^{p-1}} \cdot$$

A conclusion from (IV.18) and the preceding work is that if as $t \to a$ one of the quantities $1/r$, U, T or $\ddot{I}$ approaches ∞, they all do.

It is impossible for all the bodies to collide simultaneously after an infinite time. To prove this statement, assume the contrary, i.e., all $r_k \to 0$ for $t \to \infty$, which implies $R(t) \to 0$ for $t \to \infty$.

$r \to 0$ implies $1/r \to \infty$, which implies $\ddot{I} \to \infty$ by (IV.18) and (IV.8). If $\ddot{I} \to \infty$, at some time $\ddot{I} > 0$. For simplicity let $\ddot{I} > A$ for some t and with $A > 0$. $\ddot{I} > A$, integrated twice, yields

(IV.19) $$I > \frac{At^2}{2} + C_1 t + C_2.$$

But (IV.19) tells us that $I \to \infty$ which implies $R \to \infty$ from (IV.16). But this is contrary to our hypothesis.

Consider the problem of fewer than n-bodies colliding after some time $t = a$. Let $n = 3$, $f(r) = \gamma r^{-2}$. Then Chazy proved (see [2]) that it is impossible for a particular pair of masses to collide as $t \to \infty$ if there exists a quantity $\delta > 0$ such that both remaining distances are always greater than or equal to δ. That is, $r_{12} \nrightarrow 0$ as $t \to \infty$ if there exists δ such that r_{23} and $r_{13} \geq \delta > 0$. (See Figure 3.) We shall prove directly an extension of this result, namely that if $n = 3$, $f(r) = \gamma r^{-2}$, then $r(t) \nrightarrow 0$ as $t \to \infty$.

PROOF. Assume $r \to 0$. This implies there exists an $r_{jk} \to 0$ for some particular j and k. If no particular r_{jk} becomes and *remains* the

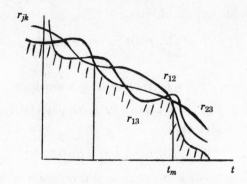

FIGURE 4. Behavior of r_{jk}

minimum r_{jk}, this implies at least two r_{jk} are alternately the minimum. Let them be r_{12} and r_{23}. Then when they exchange positions, i.e., $r_{12} < r_{23} \to r_{23} < r_{12}$, there exists a time t_m such that $r_{12} = r_{23} = r$. With an infinite number of interchanges, we have $r_{12} = r_{23} = r$ an infinite number of times. However, by Figure 3, if $r_{12} = r_{23} = r$, then $r_{13} \leq r_{12} + r_{23} = 2r$. So $R \leq 2r$ an infinite number of times. However, the reasoning used to derive (IV.19) tells us that if $r \to 0$ then $R \to \infty$. Then we cannot have $R \leq 2r$ indefinitely often as $r \to 0$. Thus if $r(t) \to 0$, a fixed r_{jk} will eventually become and remain the $r(t)$ of our definition.

From (IV.15) and previous results such as (IV.19), assuming it is r_{12} that becomes the minimum so $r_{12} \to 0$, we have as $r \to 0(1/2M)$ $(m_{23}r_{23}^2 + m_{13}r_{13}^2) \geq At^2$. Assume $m_{23} \geq m_{13}$. Then

$$(IV.20) \qquad m_{23}(r_{23}^2 + r_{13}^2) \geq At^2, \qquad r_{23}^2 + r_{13}^2 > Bt^2$$

Now let us show both r_{23} and r_{13} are greater than some multiple of t. Suppose t is large. Suppose $r_{23} \geq r_{13}$. Then by (IV.20) $r_{23} \geq (\sqrt{(B/2)})t$. As $r_{23} - r_{13} \leq r_{12} \to 0$, we can conclude $r_{13} \geq \sqrt{(B)}t/2$. If $r_{13} \geq r_{23}$, we can make a similar argument. So for large t

$$(IV.21) \qquad r_{23} \geq \sqrt{(B)}t/2, \qquad r_{13} \geq \sqrt{(B)}t/2.$$

From

$$m_3 \ddot{\mathbf{r}}_3 = \frac{\gamma m_1 m_3}{r_{13}^3}(\mathbf{r}_1 - \mathbf{r}_3) + \frac{\gamma m_2 m_3}{r_{23}^3}(\mathbf{r}_2 - \mathbf{r}_3),$$

$$(IV.22)$$

$$|\ddot{\mathbf{r}}_3| \leq \frac{\gamma m_1}{r_{13}^2} + \frac{\gamma m_2}{r_{23}^2} \leq \frac{C}{t^2}.$$

Integrate (IV.22) between $t = t_1$ and $t = t_2$, $t_1 < t_2$.

$$\left| \int_{t_1}^{t_2} \ddot{\mathbf{r}}_3 \, dt \right| \leq \int_{t_1}^{t_2} |\ddot{\mathbf{r}}_3| \, dt \leq C \left(\frac{1}{t_1} - \frac{1}{t_2} \right) ,$$

or

(IV.23)
$$|\mathbf{v}_3(t_2) - \mathbf{v}_3(t_1)| \leq C \left(\frac{1}{t_1} - \frac{1}{t_2} \right) .$$

So $\mathbf{v}_3(t_2) - \mathbf{v}_3(t_1) \to 0$ as t_1 and $t_2 \to \infty$. However, Cauchy proved that if there exists an $f(t)$, $t > 0$, such that $f(t_1) - f(t_2) \to 0$ as t_1 and $t_2 \to \infty$, then $\lim_{t \to \infty} f(t)$ exists. Applying this to $\mathbf{v}_3(t)$, $\lim_{t \to \infty} \mathbf{v}_3(\infty)$ exists.

Now let $t_2 \to \infty$, and $t_1 = t$. Then (IV.23) gives for large t

(IV.24)
$$|\mathbf{v}_3(\infty) - \mathbf{v}_3(t)| \leq \frac{C}{t} .$$

Integrating (IV.24) gives

$$\left| \int_1^t [\mathbf{v}_3(t) - \mathbf{v}_3(\infty)] \, dt \right| \leq \int_1^t |\mathbf{v}_3(t) - \mathbf{v}_3(\infty)| \, dt \leq C \log t.$$

(IV.25)
$$|\mathbf{r}_3(t) - \mathbf{v}_3(\infty) t + C_1| \leq C \log t.$$

Divide (IV.25) by t and let $t \to \infty$. Since $\log t / t$ and $c/t \to 0$,

(IV.26)
$$\lim_{t \to \infty} \frac{\mathbf{r}_3(t)}{t} = \mathbf{v}_3(\infty).$$

By (III.4) $\sum_i (m_i \mathbf{r}_i / t) = 0$, and so

(IV.27)
$$\lim_{t \to \infty} \frac{m_1 \mathbf{r}_1 + m_2 \mathbf{r}_2}{t} = \lim_{t \to \infty} \frac{-m_3 \mathbf{r}_3}{t} = -m_3 \mathbf{v}_3(\infty).$$

Recalling $\mathbf{r}_2 - \mathbf{r}_1 \to 0$ and multiplying by m_1/t,

(IV.28)
$$m_1 \frac{\mathbf{r}_2}{t} - m_1 \frac{\mathbf{r}_1}{t} \to 0.$$

Combining (IV.27) and (IV.28) we now conclude that $(m_1 + m_2)$ $(\mathbf{r}_2/t)$, (r_1/t) and (r_2/t) all have limits. Since

$$\frac{I}{t^2} = \frac{m_1 r_1^2}{t^2} + \frac{m_2 r_2^2}{t^2} + \frac{m_3 r_3^2}{t^2} ,$$

we now have $\lim_{t\to\infty} I/t^2$ exists and is finite. Dividing by t^2 and letting $t\to\infty$, (IV.19) becomes

$$\lim_{t\to\infty} \frac{I}{t^2} > \frac{A}{2}.$$

However, in (IV.19) we could take A greater than any given quantity. This gives a contradiction. Therefore $r\nrightarrow 0$ as $t\to\infty$.

In the n-body problem, a simultaneous collision of all n bodies implies that the total angular momentum is zero ($\mathbf{h} = 0$).

PROOF. Such a collision implies $R(t)\to 0$. We have previously shown $R\nrightarrow 0$ after an infinite time, so there exists a time $t = t_1 < \infty$ at which the collision must occur.

According to (III.8)

$$\mathbf{h} = \sum_{k=1}^{n} m_k(\mathbf{r}_k \times \mathbf{v}_k),$$

so that

(IV.29) $h \leq \sum_k m_k r_k v_k.$

The Cauchy inequality states that

$$\left| \sum ab \right|^2 \leq \left(\sum a^2 \right)\left(\sum b^2 \right),$$

and since (IV.29) can be written as

$$h \leq \sum_k \left((\sqrt{(m_k)}\, r_k)(\sqrt{(m_k)}\, v_k) \right),$$

we conclude

(IV.30) $h^2 \leq \left(\sum_k m_k r_k^2 \right)\left(\sum_k m_k v_k^2 \right) = 4IT.$

Using $T = \ddot{I} - E$, from (IV.10), we change (IV.30) to

(IV.31) $h^2 \leq 4I(\ddot{I} - E).$

But at some time t_1, $I\to 0$. Then by (IV.17) $R\to 0$, and so $r\to 0$. Therefore, $\ddot{I}\to\infty$, which implies that at some time $\ddot{I} > A > 0$. So the plot of I against t must be concave upwards. But this means $\dot{I} < 0$, or $-\dot{I} > 0$. Multiplying (IV.31) by $-\dot{I}/I$,

$$\frac{h^2(-\dot{I})}{I} \leq -4\dot{I}(\ddot{I} - E).$$

Integrating this in the neighborhood of t_1,

(IV.32) $h^2 \log \dfrac{1}{I} \leqq 4EI - 2\dot{I}^2 + K \leqq 4EI + K.$

As $1/I \to \infty$, $\log(1/I) \to \infty$, and so is eventually > 0. Dividing (IV.32) by $\log(1/I)$,

$$h^2 \leqq \frac{4EI + K}{\log \dfrac{1}{I}} \text{ for } t \sim t_1.$$

Now as $t \to t_1$, the denomiator $\to \infty$, $I \to 0$ and the numerator $\to K$. Thus $h^2 \to 0$ as $t \to t_1$, or the total angular momentum vanishes if there exists a simultaneous collision of all masses.

V. A Tauberian Theorem and its application. Consider the problem of a given function $f(x)$, $x > 0$ such that $f(x) \sim Ax^2$, i.e., $\lim_{x \to \infty} f(x)/x^2 = A$, and let us question if it is true that $f'(x) \sim 2Ax$.

The converse statement (that if $f'(x) \sim 2Ax$, then $f(x) \sim Ax^2$) is true, but it is not necessarily true that given an asymptotic relation one can differentiate with the result being also an asymptotic relation. It is this irreversibility that led to the concept of the Tauberian condition, which is that additional information required to obtain reversibility in the above limits.

We first prove that if $f'(x) \sim 2Ax$, then $f(x) \sim Ax^2$. If $\lim_{x \to \infty} f'(x)/2x = A$, then for each $\epsilon > 0$ there exists an x_0 such that

(V.1) $\left| \dfrac{f'(x)}{2x} - A \right| < \epsilon \text{ for } x \geqq x_0.$

Multiply (V.1) by $2x$, integrate with respect to x, divide by x^2 and let $x \to \infty$.

(V.2) $\left| \dfrac{f(x)}{x^2} - A \right| \leqq \epsilon + \dfrac{C}{x^2}$

for some constant C. Since ϵ is arbitrary, let $\epsilon \to 0$ so (V.2) leads to

(V.3) $\lim\limits_{x \to \infty} \dfrac{f(x)}{x^2} = A \text{ or } f(x) \sim Ax^2.$

Now, to show that if $f(x) \sim Ax^2$ then $f'(x) \sim 2Ax$ we must introduce a Tauberian condition.

LANDAU'S THEOREM (1906). *If $f(x) \sim Ax^2$, and $f''(x) \geq C > -\infty$, then $f'(x) \sim 2Ax$.*

Case 1. $A = 0$. Consider $f(x + \epsilon x)$, and its expansion in Taylor series with a remainder, up to second-order terms.

$$(V.4) \qquad f(x + \epsilon x) = f(x) + \epsilon x f'(x) + \frac{\epsilon^2 x^2}{2} f''(\xi),$$

where ξ lies between x and $x + \epsilon x$. Thus

$$(V.5) \qquad f(x + \epsilon x) \geq f(x) + \epsilon x f'(x) + \frac{\epsilon^2 x^2}{2} \cdot C.$$

Dividing (V.5) by x^2 and taking x large enough so that

$$\left| \frac{f(x)}{x^2} - A \right| \leq \epsilon^2 \text{ and } \left| \frac{f(x + \epsilon x)}{(x + \epsilon x)^2} - A \right| \leq \epsilon^2,$$

and recalling that $A = 0$, we have

$$(V.6) \qquad \frac{\epsilon f'(x)}{x} \leq \epsilon^2 \left(1 + (1 + |\epsilon|)^2 - \frac{C}{2} \right).$$

In (V.6), for $\epsilon > 0$, divide by ϵ

$$(V.7) \qquad \frac{f'(x)}{x} \leq \epsilon \left(1 + (1 + |\epsilon|)^2 - \frac{C}{2} \right).$$

In (V.6), for $\epsilon < 0$, divide by ϵ (reverse inequality) and (V.6) becomes

$$(V.8) \qquad \frac{f'(x)}{x} \geq \epsilon \left(1 + (1 + |\epsilon|)^2 - \frac{C}{2} \right).$$

Clearly (V.7) and (V.8) imply

$$\lim_{x \to \infty} \frac{f'(x)}{2x} = 0.$$

Case 2. $A \neq 0$. Define $g(x)$ such that

$$(V.9) \qquad \frac{g(x)}{x^2} = \frac{f(x)}{x^2} - A.$$

The hypothesis $f(x) \sim Ax^2$ implies $g(x)/x^2 \to 0$. From (V.9), $g''(x) = f''(x) - 2A$.

If $f''(x) \geq C > - \infty$, then $g''(x) \geq C - 2A > - \infty$. Hence the argument for the case $A = 0$ now applies to $g(x)$, so

(V.10) $$\lim_{x \to \infty} \frac{g'(x)}{2x} = 0.$$

By (V.9),

(V.11) $$\frac{g'(x)}{2x} = \frac{f'(x)}{2x} - A.$$

Using (V.10) and (V.11), $f'(x) \sim 2Ax$.

The same conclusion can be reached for $f''(x) \leq C < \infty$ by using $- f$ for f in the above argument.

VON CLAUSIUS' THEOREM. *If we assume the inverse square law, and a system is bounded in size and velocity, then both the kinetic energy and potential energy have averages in the limit sense. Specifically*

(V.12) $$\hat{U} = \lim_{t \to \infty} \frac{1}{t} \int_0^t U \, dt = - 2E,$$

(V.13) $$\hat{T} = \lim_{t \to \infty} \frac{1}{t} \int_0^t T \, dt = - E.$$

PROOF. If (V.12) is true, (V.13) follows, since $T = U + E$ by (III.11), and since E is constant it is equal to its average value. Thus $\hat{T} = \hat{U} + E$. It other words, (V.13) is redundant in view of the conservation of energy.

Now to establish (V.12). Begin with the Lagrange-Jacobi identity $\ddot{I} = U + 2E$, integrate once and divide by t.

(V.14) $$\frac{\dot{I}}{t} = \frac{1}{t} \int_0^t U \, dt + 2E + \frac{K}{t}.$$

Remembering (IV.2), from the hypothesis of bounded $\mathbf{r}_k$ and $\mathbf{v}_k$, $\dot{I}$ is bounded in time. Thus, as $t \to \infty$, $\dot{I}/t \to 0$, $K/t \to 0$, and

$$\hat{U} = \lim_{t \to \infty} \frac{1}{t} \int_0^t U \, dt = - 2E.$$

Due to the fact that in some cases, e.g., the parabolic case of the two body problem, $\hat{U} = - 2E (= 0)$ even for an unbounded system $(r \sim t^{2/3})$, it is possible to prove a stronger theorem.

THEOREM. *A necessary and sufficient condition that $\hat{U}$ exist and equal $-2E$ is that $\lim_{t \to \infty} R(t)/t = 0$.*

Here no condition is put upon the velocity, and the system can grow with time provided it grows at a slower rate than t.

To prove the stated theorem, let us first show that $\dot{I}/t \to 0$ if and only if $I/t^2 \to 0$. The argument given before our proof of Landau's Theorem shows that if $\dot{I}/t \to 0$, then $I/t^2 \to 0$.

We can get the reverse by Landau's Theorem if we can show $I > C > -\infty$. However, by the Lagrange-Jacobi identity $\ddot{I} = U + 2E > 2E > -\infty$.

By (V.14), $\hat{U} = -2E$ if and only if $\dot{I}/t \to 0$. So $\hat{U} = -2E$ if and only if $I/t^2 \to 0$. Then by (IV.17), $\hat{U} = -2E$ if and only if $R(t)/t \to 0$. Thus our theorem is proved.

THEOREM. *If $\hat{T}$ exists and equals 0, then $E = 0$.*

PROOF. $T = U + E$, so if $\hat{T}$ exists, so does $\hat{U}$, and $\hat{T} = \hat{U} + E$. As $\hat{T} = 0$ by hypothesis, $\hat{U} = -E$. But $\hat{U} \geq 0$, so $-E \geq 0$ or

$$(V.15) \qquad\qquad\qquad E \leq 0.$$

From $\ddot{I} = T + E$, integration once and division by t gives

$$(V.16) \qquad\qquad \frac{\dot{I}}{t} = \frac{1}{t} \int_0^t T \, dt + E + \frac{C}{t}.$$

As $t \to \infty$, if $\hat{T} = 0$, the integral in (V.16) must vanish, as does C/t. Thus:

$$\lim_{t \to \infty} \frac{\dot{I}}{t} = E \text{ so } \dot{I} \sim Et.$$

Integration of this asymptotic function gives

$$I \sim \frac{Et^2}{2},$$

or

$$\lim_{t \to \infty} \frac{I}{t^2} = \frac{E}{2}.$$

But $I > 0$, so $E \geq 0$. Combining this result with (V.15), we have $E = 0$.

Incidentally, the theorem is true for all $p \neq 1$.

What happens if $E < 0$? Since $T = U + E$, and $T \geq 0$, by hypothesis

$$U + E \geq 0 \quad \text{or} \quad U \geq -E.$$

Thus $U \geq |E|$, but $A/r \geq U$ by (IV.18), so $A/r \geq |E|$. Thus $A/|E| \geq r$. So if $E < 0$, then r is bounded but R could conceivably $\rightarrow \infty$.

If $E > 0$, there are some interesting conclusions to be drawn.

THEOREM. *If $E > 0$, $n = 3$, a particle escapes.*

LEMMA 1. $4EI - \dot{I}^2 < C$ *as* $t \rightarrow \infty$.

From the Lagrange-Jacobi equation $\ddot{I} = 2E + U$, we conclude $\dot{I} \rightarrow \infty$ as $t \rightarrow \infty$. Thus, for some time $a\dot{I}$ becomes positive, say $\dot{I}(a) = k > 0$. By integrating $\ddot{I}$,

(V.17) $$\dot{I} = \int_a^t U(\tau)d\tau + 2E(t - a) + k.$$

(V.18) $$I = \int_a^t (t - \tau) U(\tau)d\tau + E(t - a)^2 + k(t - a) + l.$$

Multiplying (V.18) by $4E$, we can write

(V.19) $$4EI < 4E(t - a) \int_a^t U(\tau)d\tau + 4E^2(t - a)^2 + 4Ek(t - a) + C.$$

Squaring both sides of (V.17) gives

(V.20) $$\dot{I}^2 > 4E^2(t - a)^2 + 4E(t - a) \int_a^t U(\tau)d\tau + 4Ek(t - a).$$

Subtracting (V.20) from (V.19) gives

$$4EI - \dot{I}^2 < C.$$

Define $J = \sum_{1 \leq j < k \leq n} m_{jk} r_{jk}.$
A familiar argument gives

(V.21) $$BR < J < AR.$$

LEMMA 2. $\lim_{t \rightarrow \infty} 1/UJ$ *exists.*

From the definitions,

$$UJ = \sum \frac{\gamma m_{jk}}{r_{jk}} \sum m_{pq} r_{pq}, \qquad 1 \leqq j < k \leqq n, \quad 1 \leqq p < q \leqq n$$

$$= \sum \sum \gamma m_{jk} m_{pq} \frac{r_{pq}}{r_{jk}}.$$

Differentiating this with respect to time,

$$(V.22) \qquad (UJ)' = \sum \sum \gamma m_{jk} m_{pq} \frac{r_{jk} \dot{r}_{pq} - r_{pq} \dot{r}_{jk}}{r_{jk}^2}$$

But $r_{jk} \geqq r$ by definition, so $1/r_{jk} \leqq 1/r$. From (V.22) and (IV.18), with $C = B^{-2}$,

$$(V.23) \quad |(UJ)'| \leqq CU^2 \sum \sum \gamma \sqrt{(m_{jk} m_{pq})} \sqrt{(m_{jk} m_{pq})} \, |r_{jk} \dot{r}_{pq} - r_{pq} \dot{r}_{jk}|.$$

Square (V.23) use Cauchy's inequality, and note that in the expansion of $|\quad|^2$ we get a middle term

$$- 2 \sum m_{jk} r_{jk} \dot{r}_{jk} \sum m_{pq} \dot{r}_{pq} = - 2M^2 \dot{I}^2.$$

Thus (V.23) becomes

$$|(UJ)'|^2 \leqq U^4 (CJ^2 U + C_1) \qquad (J^2 > I^2).$$

Dividing by $U^5 J^5$,

$$(V.24) \qquad \frac{|(UJ)'|^2}{U^5 J^5} \leqq C \cdot \frac{1}{J^3} + \frac{C_1}{J^4} \cdot \frac{1}{UJ}.$$

But by (IV.18) and (V.21) $1/U \leqq Cr$ and $1/J \leqq C/R$. Thus $1/UJ \leqq Cr/R \leqq C$. This implies $1/UJ$ is bounded, and it will be dropped from the right side of (V.24). Similarly, since $E > 0$, if $I > At^2$, then $J^2 > At^2$, and so $J > At$. Finally (V.24) can be reduced to

$$\frac{|(UJ)'|}{(UJ)^{5/2}} \leqq \frac{B}{t^{3/2}}.$$

Integration of this gives

$$(V.25) \qquad \left| \frac{1}{(UJ)_1^{3/2}} - \frac{1}{(UJ)_2^{3/2}} \right| \leqq B \left| \frac{1}{t_1^{1/2}} - \frac{1}{t_2^{1/2}} \right|.$$

But as t_1 and $t_2 \to \infty$ independently, the right side of (V.25) vanishes, and $1/(UJ)^{3/2} \to \lim$ as $t \to \infty$ or $\lim_{t \to \infty} 1/UJ = l$.

Assume $l > 0$. Then, there exists a $\delta > 0$ such that $1/UJ \geqq \delta > 0$ as $t \to \infty$,

(V.26) $$\frac{r}{R} \geqq \frac{C}{UJ} \geqq \delta > 0, \qquad \text{so } r \geqq \delta R \geqq Ct,$$

(V.26) shows that as $t \to \infty$, $r \to \infty$ so all the particles escape.

Assume $l = 0$. Then at some time, some r_{jk}(say r_{12}) becomes the minimum, i.e., $r_{12} = r$. If another r_{jk} swaps with r_{12}, each time a new r_{jk} becomes the minimum, $r_{12} = r_{23}$. (Assuming r_{23} is the other minimum.) Then $r_{13} = R$. But

$$|r_{12} - r_{13}| \leqq r_{23},$$

(V.27)   or

$$|r - R| \leqq r.$$

Division of (V.27) by R, gives $|r/R - 1| \leqq r/R$. But $\lim_{t \to \infty} r/R = 0$, which implies $|-1| \leqq 0$. Therefore there exists a min r_{jk}; call it r_{12}.

Introduce Jacobi coordinates ξ and r, where

$$\xi = \mathbf{r}_3 - \frac{m_1 \mathbf{r}_1 + m_2 \mathbf{r}_2}{m_1 + m_2},$$

$$\mathbf{r} = \mathbf{r}_2 - \mathbf{r}_1,$$

$$m_1 \mathbf{r}_1 + m_2 \mathbf{r}_2 + m_3 \mathbf{r}_3 = 0.$$

The above set of equations may be solved to get $\mathbf{r}_1$, $\mathbf{r}_2$ and $\mathbf{r}_3$ as linear functions of $\mathbf{r}$ and ξ. Such a manipulation would show

(V.28) $$I = A\xi^2 + Br^2,$$

where A and B are functions only of the masses.

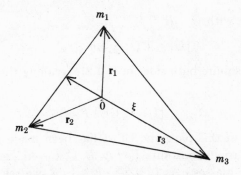

FIGURE 5. Jacobi Coordinates

Let $r_{23} = R$, $r_{12} = r$. Then $|R - \xi| \leq r$, or $|1 - (\xi/R)| \leq r/R$. However, by (IV.18) and (V.21), $1/U \geq Cr$ and $1/J \geq C/R$. So $r/R \leq C/UJ$. With $l = 0$, this gives $\xi/R \to 1$, $\xi \sim R$.

Since $R \geq Ct$, $\xi \sim Ct$ so ξ becomes unbounded as $t \to \infty$. But rewriting the form for ξ

$$\xi = \frac{M\mathbf{r}_3 - \sum m_i \mathbf{r}_i}{M - m_3} = \frac{M}{M - m_3}\mathbf{r}_3 = c\mathbf{r}_3.$$

Since, as $t \to \infty$, $\xi \to \infty$, we conclude $r_3 \to \infty$. This means m_3 escapes from the system.

VI. **Classification of motions.** Recall the general equation of motion

(VI.1) $$m_k \ddot{\mathbf{r}}_k = \sum_{1 \leq j < k \leq n} m_{jk} \frac{f(r_{jk})}{r_{jk}} (\mathbf{r}_j - \mathbf{r}_k).$$

When $f(r_{jk})$ is a real, analytic function there exists a unique set of $r_k(t)$ which satisfies (VI.1). For the case where $f(r) = \gamma/r^2$, either all the $r_k(t)$ may be continued analytically as $t \to \infty$, or there is some time $t = t_1$ at which at least one r_k ceases to be analytic. Painlevé has shown (see [3]) that the solution of the n-body problem permits analytic continuation until such time $t = t_1$ for which $r(t) \to 0$ as $t \to t_1$, and that this condition is both necessary and sufficient.

Returning to the work of Chazy in [2] for the case where $t \to \infty$, is it possible to find estimates for the growth in $r(t)$ with time?

From the definition of U,

$$-\dot{U} = \sum \frac{\gamma m_{jk} \dot{r}_{jk}}{r_{jk}^2}.$$

From (IV.18), with $C = B^{-2}$,

(VI.2) $$|\dot{U}| \leq CU^2 \sum \gamma m_{jk} |\dot{r}_{jk}|.$$

As before, we square both sides of (VI.2), recalling the energy equations, to get

(VI.3) $$|\dot{U}|^2 \leq CU^4 T.$$

(Note that in this paper no effort has been made to distinguish between the various constants, i.e., A, B, C, C_1, etc., since they only depend on the masses.) As $U = T - E \geq -E$, we see that if $E < 0$ then

$$\frac{A}{r} \geqq U \geqq |E|,$$

(VI.4)
$$r \leqq \frac{A}{|E|}.$$

If $E = 0$, $T = U$ and (VI.3) becomes

$$|\dot{U}|^2 \leqq CU^5,$$

(VI.5)
$$\frac{|\dot{U}|}{U^{5/2}} \leqq B.$$

Integration gives $U^{-3/2} \leqq B_1 t + B_2$, so that $U^{-1} \leqq Bt^{2/3}$ for large t. As $A/r \geqq U$,

(VI.6)
$$r \leqq Ct^{2/3}.$$

If $E > 0$, (VI.3) becomes

(VI.7)
$$|\dot{U}|^2 \leqq CU^4(U + E).$$

Let $\xi = U^{-1}$, so that from (VI.7) we get

$$|\dot{\xi}\xi^{-2}|^2 \leqq C\xi^{-4}(\xi^{-1} + E)$$

(VI.8)
$$|\dot{\xi}| \leqq \sqrt{\left(\frac{C + CE\xi}{\xi}\right)}.$$

Choose A large enough so that

(VI.9)
$$A \geqq 2\sqrt{\left(\frac{C + CEA}{A}\right)}.$$

We wish to show that for sufficiently large t_0

(VI.10)
$$\xi \leqq At$$

for all $t \geqq t_0$. We first suppose that for all $t \geqq t_1$, $\xi \geqq A$. Then by (VI.8) and (VI.9),

(VI.11)
$$|\dot{\xi}| \leqq \frac{1}{2} A$$

so that (VI.10) will eventually hold. Alternatively suppose that for every t_1 there is a greater t at which $\xi < A$. Take t_0 to be such a $t > 1$. Then we prove (VI.10) by reductio ad absurdum. Suppose (VI.10) fails. As $t_0 > 1$ and $\xi < A$ at $t = t_0$, we have (VI.10) holding at $t = t_0$, and consequently for $t_0 \leq t \leq t_2$, where t_2 is the greatest lower bound of all t's greater than t_0 for which (VI.10) fails. Then $\xi \geq At_2$ at $t = t_2$, and so (since $t_2 > t_0 > 1$), we must have $\xi \geq A$ for $t_2 - \epsilon \leq t \leq t_2$. Then by (VI.8) and (VI.9), (VI.11) must hold for $t_2 - \epsilon \leq t \leq t_2$. With $\xi \geq At_2$ at $t = t_2$, this tells us that $\xi > At$ at $t = t_2 - \epsilon$, contradicting our definition of t_2 as the greatest lower bound of t's greater that t_0 for which (VI.10) fails.

From (VI.10), we get $Cr \leq U^{-1} = \xi \leq At$.

Thus

(VI.12) $$r \sim t.$$

For $n = 3$, and ruling out the case $h = 0$ of triple collision and writing $l = \lim_{t \to \infty} (UJ)^{-1}$, Chazy gave the following possibilities (see [2]):

$E > 0$	all	$r_{ij} \sim t$	$l \neq 0$	hyperbolic case
	two	r_{ij}'s, say r_{12}		
	and	$r_{23} \sim t$	$l = 0$	hyperbolic-parabolic case
		$r_{13} \sim t^{2/3}$		
	two	r_{ij}'s, say r_{12}		
	and	$r_{23} \sim t$	$l = 0$	hyperbolic-elliptic case
		$r_{13} \leq B$		
$E = 0$	two	r_{ij}'s, say r_{12}		
	and	$r_{23} \sim t$		hyperbolic-elliptic case
		$r_{13} \leq B$		
	all	$r_{ij} \sim t^{2/3}$		parabolic case
$E < 0$	All the above cases. In addition a case in which some r_{ij} oscillates infinity.			

Chazy was able to exhibit orbits of every type above save for the very last, but the latest Russian literature indicates this too has now been exhibited.

Let us now return to the Sundman problem, namely the three-body problem where $h \neq 0$, and show that if $r \to 0$ as $t \to t_1$, this corresponds to a two particle collision with the third particle moving to a definite position with a definite velocity.

If $r(t) \to 0$ as $t \to t_1$, $U \to \infty$ so $\ddot{I} \to \infty$. But this tells us that the curve of I against t must be concave upward in the neighborhood of $t = t_1$, so $I \to L$, where $0 \leq L \leq \infty$. We have already ruled out the case where $I = 0$, so we have $0 < L \leq \infty$.

If $r \to 0$, a pair of particles collide. Then one of the distances r_{ij} becomes a minimum and remains so. To prove this, assume r_{12} and r_{23} are alternately the minimum. Then there exists a sequence of times $\{t_n\}$, $t_n \to t_1$ where $r_{12}(t_n) = r_{23}(t_n) \to 0$. But $r_{13} \leq r_{12} + r_{23}$, so $r_{13}(t_n) \to 0$ which implies all three $r_{ij} \to 0$ along this sequence of t_n's. But note that by (IV.15) $I = 1/2M \sum m_{jk} r_{jk}^2$, so $I(t_n) \to 0$ along this sequence.

We have already proved I has a limit, so if it approaches 0 along a particular sequence of times, it will approach 0 no matter how you approach t_1. But this would be a triple collision, contrary to $\mathbf{h} \neq 0$. Thus only one r_{ij} eventually becomes and stays the minimum.

Now to show that the third particle moves to a definite position with a definite velocity. If $r_{12} \to 0$, then $r_{23} - r_{13} \to 0$. We know $m_{12} r_{12}^2 \to 0$. This implies by (IV.15) that $m_{23} r_{23}^2 + m_{31} r_{13}^2 \to 2ML > 0$.

As $r_{23} - r_{13} \to 0$, we must have both r_{23} and r_{13} approaching the same finite limit, which must be positive, or going to ∞ together. But

$$\ddot{\mathbf{r}}_3 = \frac{\gamma m_1}{r_{13}^3} (\mathbf{r}_1 - \mathbf{r}_3) + \frac{\gamma m_2}{r_{23}^3} (\mathbf{r}_2 - \mathbf{r}_3),$$

so

$$|\ddot{\mathbf{r}}_3| \leq \frac{\gamma m_1}{r_{13}^2} + \frac{\gamma m_2}{r_{23}^2}$$

As r_{23} and r_{13} are bounded away from zero, we see that $\ddot{\mathbf{r}}_3$ remains bounded. Recalling that if a function has a bounded derivative as $t \to t_1$, the function itself has a limit, we see that $\dot{\mathbf{r}}_3$ has a limit as $t \to t_1$. So $\dot{\mathbf{r}}_3$ is bounded, and $\mathbf{r}_3 \to$ limit as $t \to t_1$.

However, $m_1 \mathbf{r}_1 + m_2 \mathbf{r}_2 + m_3 \mathbf{r}_3 = 0$, so that $m_1 \mathbf{r}_1 + m_2 \mathbf{r}_2$ has a limit. Also, $r_{12} \to 0$, $\mathbf{r}_1 - \mathbf{r}_2 \to 0$, from which we conclude that $\mathbf{r}_1$ has a limit, which is the same limit that $\mathbf{r}_2$ has. That is, $\mathbf{r}_1$, $\mathbf{r}_2$, and $\mathbf{r}_3$ all have limits, which says that the particles approach definite positions. Also, by (IV.1), we conclude that L is finite.

VII. **The Sundman problem.** Among the theorems proved by Sundman (see [4]) are the following, where $n = 3$, we have chosen the units so that $f(r) = r^{-2}$, $h \neq 0$, and $r(t) \to 0$ as $t \to t_1 < \infty$.

THEOREM 1. $\lim_{t \to t_1} I(t) = L$, $0 < L < \infty$.

THEOREM 2. *Two particles collide and the third particle goes to a definite position with a finite velocity.*

THEOREM 3. *If $\mathbf{v}_1$ and $\mathbf{v}_2$ are the velocities of the colliding particles, then*

$$\lim_{t \to t_1} r(t) v_1^2 = 2m_2^2/(m_1 + m_2),$$

$$\lim_{t \to t_1} r(t) v_2^2 = 2m_1^2/(m_1 + m_2).$$

THEOREM 4. *The integral*

$$\int^t U(\tau) d\tau, \text{ converges } (cf. \ h = 0 \text{ in two-body case}).$$

THEOREM 5. *If $\mathbf{V}$ is the velocity of m_1 relative to m_2, then at collision*

$$\lim_{t \to t_1} r V^2 = 2(m_1 + m_2).$$

THEOREM 6. $\lim_{t \to t_1} r(d^2/dt^2)(r^2) = 2(m_1 + m_2)$.

THEOREM 7. *As $t \to t_1$, $r \sim (t_1 - t)^{2/3}$.*

We undertake to prove these. We have already proved Theorems 1 and 2. As in their proofs, we take $\mathbf{r} = \mathbf{r}_2 - \mathbf{r}_1$. By definition

$$U = \frac{m_{12}}{r_{12}} + \frac{m_{23}}{r_{23}} + \frac{m_{31}}{r_{31}}.$$

Multiply this by r and let $t \to t_1$, to obtain

(VII.1) $$\lim_{t \to t_1} rU = m_{12}.$$

From this

$$r(T - E) \to m_{12}.$$

But E is constant, $r \to 0$, so we can write $rT \to m_{12}$, or

$$r(m_1 v_1^2 + m_2 v_2^2 + m_3 v_3^2) \to 2m_{12}.$$

However, V_3 approaches a limit, so that

(VII.2) $$r(m_1 v_1^2) + r(m_2 v_2^2) \to 2m_{12}.$$

The center of mass is such that

(VII.3) $$m_1 \mathbf{v}_1 + m_2 \mathbf{v}_2 + m_3 \mathbf{v}_3 = 0.$$

Let us multiply (VII.3) by $\sqrt{r}$, recall that $\mathbf{v}_3$ approaches a limit $\sqrt{r} \to 0$, and combine (VII.2) and (VII.3) to get

(VII.4) $$rv_1^2 \to \frac{2m_2^2}{m_1 + m_2},$$

(VII.5) $$rv_2^2 \to \frac{2m_1^2}{m_1 + m_2}.$$

From (VII.3) we get

$$rm_1 \mathbf{v}_1^2 + rm_2 \mathbf{v}_1 \cdot \mathbf{v}_2 + rm_3 \mathbf{v}_1 \cdot \mathbf{v}_3 = 0.$$

As $\sqrt{(r)}\mathbf{v}_1$ is bounded by (VII.4) and v_3 approaches a limit, we conclude by (VII.4)

$$r\mathbf{v}_1 \cdot \mathbf{v}_2 \to \frac{-2m_1 m_2}{m_1 + m_2}.$$

So by (VII.4) and (VII.5)

(VII.6) $$r|\mathbf{v}_1 - \mathbf{v}_2|^2 \to 2(m_1 + m_2).$$

Theorems 3 and 5 follow from (VII.4), (VII.5), and (VII.6). By (IV.18), Theorem 4 will follow if we can show

$$\lim_{t \to t_1} \int^t \frac{d\tau}{r(\tau)} \text{ is bounded.}$$

Since $1/r \sim U = \ddot{I} - 2E$, $\ddot{I} \to \infty$ as $t \to t_1$, so at some point $\ddot{I}$ becomes positive. Thus, if we show the existence of a limit for $\int^t \ddot{I} d\tau$, we have a limit for $\int^t d\tau/r(\tau)$. Now, by (IV.1)

$$\dot{I} = m_1(\mathbf{r}_1 \cdot \mathbf{v}_1) + m_2(\mathbf{r}_2 \cdot \mathbf{v}_2) + m_3(\mathbf{r}_3 \cdot \mathbf{v}_3)$$

$$= \mathbf{r}_1(m_1 \mathbf{v}_1 + m_2 \mathbf{v}_2) + m_2(\mathbf{r}_2 - \mathbf{r}_1)\mathbf{v}_2 + m_3 \mathbf{r}_3 \cdot \mathbf{v}_3.$$

So by (III.5)

$$\dot{I} = m_2 \mathbf{r} \cdot \mathbf{v}_2 + m_3 (\mathbf{r}_3 - \mathbf{r}_1) \cdot \mathbf{v}_3.$$

We showed in §VI that $\mathbf{r}_1, \mathbf{r}_3$, and $\mathbf{v}_3$ have limits. Also

$$|\mathbf{r} \cdot \mathbf{v}_2| \leqq \sqrt{r} \sqrt{r}\, v_2$$

so that $\mathbf{r} \cdot \mathbf{v}_2 \to 0$ by (VII.5). This proves Theorem 4.

We have

(VII.7) $$\frac{d^2}{dt^2}(r^2) = 2\mathbf{r} \cdot \ddot{\mathbf{r}} + 2\dot{r}^2.$$

By (VI.1)

$$\ddot{\mathbf{r}}_1 = \frac{m_2}{r_{12}^3}(\mathbf{r}_2 - \mathbf{r}_1) + \text{bounded term},$$

$$\ddot{\mathbf{r}}_2 = \frac{m_1}{r_{12}^3}(\mathbf{r}_1 - \mathbf{r}_2) + \text{bounded term}.$$

Subtract the first from the second,

$$\ddot{\mathbf{r}} = -\frac{m_1 + m_2}{r^3}\mathbf{r} + \text{bounded term}.$$

So,

$$\mathbf{r} \cdot \ddot{\mathbf{r}} = -\frac{m_1 + m_2}{r} + \text{vanishing term}.$$

Using this with (VII.6) and (VII.7) gives Theorem 6.

To find the limit of $r\dot{r}^2$, we write

$$r\dot{r}^2 = F = \frac{r^2 \dot{r}^2}{r}$$

and use l'Hôpital's rule.

$$\lim_{t \to t_1} r\dot{r}^2 = \lim_{t \to t_1} \frac{2r\dot{r}^3 + 2r^2\dot{r}\ddot{r}}{\dot{r}} = \lim_{t \to t_1} 2r(\dot{r}^2 + r\ddot{r})$$

$$= \lim_{t \to t_1} r\frac{d^2}{dt^2}(r^2) = 2(m_1 + m_2)$$

by Theorem 6. Thus

(VII.8) $$\sqrt{r}\,\dot{r} \to \pm \sqrt{(2(m_1 + m_2))}.$$

But $r \to 0$, so (VII.8) becomes

$$\sqrt{r\dot{r}} \to -\sqrt{(2(m_1 + m_2))},$$

and

$$\int_r^0 \sqrt{r}\,dr \sim -\sqrt{(2(m_1 + m_2))}\,(t_1 - t).$$

Integrating the left side of this gives

$$\frac{2}{3}\,r^{3/2} \sim \sqrt{(2(m_1 + m_2))}\,(t_1 - t).$$

So Theorem 7 holds.

When two particles approach collision, the forces between them approach infinity. When $h \neq 0$, so that the third particle remains at a finite distance the forces which it exerts on the two colliding particles remain finite; hence they become negligible compared to the force between the colliding particles. Thus if two particles collide (or come extremely close together) the local behavior of the two particles should approximate that of a pure two-body problem.

Theorems 3 through 7 verify particular aspects of this. Sundman (see [4]) verified still other aspects, such as that the colliding particles approach at a definite angle. However, this is a purely mathematical result, based on the assumption that the particles are dimensionless points which collide by coming into exact coincidence. The chance that this might occur with real objects of finite dimension is vanishingly small.

If two dimensionless particles have a very near miss, each will approach the other, run around it, and recede on what is nearly a hyperbola (approximating the exact hyperbolas of the pure two-body problem). If the particles have appreciable size, this would be a collision; this could vary from nearly head on to grazing impact, depending on the closeness of approach in the idealized case of dimensionless particles.

References

1. J. Bertrand, C. R. Acad. Sci. Paris **77**(1873), 849-853.
2. J. Chazy, Ann. Sci. École Norm. Sup. **39**(1922), 124.
3. P. Painlevé, C. R. Acad Sci. Paris **123**(1896), 636-639; 871-873.
4. K. F. Sundman, Acta. Math. **36**(1913), 105.

PURDUE UNIVERSITY

Index